Cognitive Science:
Recent Advances and Recurring Problems

Frederick Adams
University of Delaware

Osvaldo Pessoa Jr.
University of Sao Paulo, Brazil

Joao Eduardo Kogler Jr.
University of Sao Paulo, Brazil

Cognitive Science & Psychology

VERNON PRESS

www.vernonpress.com

In the Americas:
Vernon Press
1000 N West Street,
Suite 1200, Wilmington,
Delaware 19801
United States

In the rest of the world:
Vernon Press
C/Sancti Espiritu 17,
Malaga, 29006
Spain

Vernon Series in Cognitive Science and Psychology

Library of Congress Control Number: 2017904399

ISBN: 978-1-62273-100-8

Table of Contents

Introduction

This book is a survey of various topics pertaining to cognitive science and philosophy of mind, as presented in an international meeting held in the city of São Paulo, the *10th Brazilian International Meeting of Cognitive Science* (EBICC 2015).

Part I of the book brings eight chapters written by the invited speakers of the meeting, with topics ranging from brain connectivity, adaptive computation, and human visual system to issues in epistemology, theory of action, philosophy of mind, and friendship. Part II presents four chapters written is essay style, addressing topics discussed in some round tables at the conference, and one of them (chapter ten) features the paper chosen as the best student essay of the conference. These chapters cover subjects spanning from spike propagation neural networks, logical and geometrical aspects of cognition, the frame problem, to ethical issues in information technology. Part III contains eight chapters, each one about a topic with recurring questions in the field. They address problems in computer simulations in cognitive science, cognitive architectures, Bayesian statistical reasoning, deep learning, long-term synaptic plasticity, language and cognition, the mind-body problem, issues in semiotics, and in the history and philosophy of neuroscience.

The book is targeted for researchers, teachers and students, particularly those interested in the fields of cognitive science, artificial intelligence, and philosophy of mind, and also covers topics of relevance in linguistics, neuroscience, biology, psychology, physics, mathematics, and computer science.

Part I opens with the contribution of Maria Eunice Gonzalez, who discusses to what extent the concept of "autonomous action", applicable to human beings, can be extended to robots and machines. She examines some recent examples of adaptive self-organizing robots, and the dangers they might pose for human autonomy. The Complex Systems Paradigm is used as a theoretical guide for analyzing the emergent properties of the multiple-scale interactions between humans and autonomous learning machines.

In the second chapter, Fred Adams discusses the issue of "content externalism" in epistemology and philosophy of language. Externalism about mental content is the position that the content of an intentional state, such as a belief or a desire, does not only supervene on the intrinsic properties of a subject, but also depends on the subject's relation to the external environment. This is Adams' position, which is also called "broad content", and he defends content externalism from Gabriel Segal's attacks against it. After presenting Segal's views on "narrow content", especially his argument based on empty terms, such as "Santa Claus", Adams argues extensively that there is very slim support for narrow content. In his radical externalist account, empty names and kind terms have no meaning because they have no bearers (objects or properties).

In chapter three, Claus Emmeche analyzes friendship as a social cognitive phenomenon, within a framework of multilevel organization, with different levels of embodied, extended and embedded cognition. Distinct levels of embodied cognition are identified as animate, anthropic and societal. Friendship is understood as involving "relational attention", which contains three aspects: self-disclosure, mutual knowledge, and joint attention (*sunaisthesis*). Friendship also involves thinking together, which expresses patience and empathy. The author has shown how philosophizing about friendship may be valuable for cognitive science.

In chapter four, Sofia Stein analyzes the mind-body problem from the perspective of neutral monism – the view that there are basic sense-data from which any object is constructed, as well as the self. Adopting a position close to science, involving a pragmatic stance and considerations from the philosophy of language, she concludes that neutral monism is in fact a good approach to the problem.

Dora Fix Ventura and coworkers discuss in chapter five some recent research findings arising from the exciting discovery made in 1998 that there is a third type of photosensitive cell in the human eye, a subset of retinal ganglion cells that contains the pigment melanopsin, maximally activated by blue light. Such cells initiate a non-image-forming visual pathway, responsible for pupillary light reflex, circadian rhythm of sleep and awakeness, mood, cognitive capacity, and playing a role in the image-forming system. The authors also review their original laboratory work exploring pupillary light reflex in patients with glaucoma and Leber's hereditary optic neuropathy.

Koichi Sameshima, with collaboration of Luiz A. Baccalá, presents in chapter six an overview of the experimental modeling of functional features of the brain, based on the analysis of correlations and causality present in electrophysiological data and functional measurements of brain activity. They point to the recent question of "how brain areas interact and integrate information (functional integration) during the enactment of cognitive tasks or over the genesis of behavior". They provide a historical and methodological review of the efforts that led to the analysis of brain connectivity, based on the use of a measure of "partial directed coherence", closely related to the temporal correlation measure known as "Granger causality". Finally, they make some remarks about the challenges posed by the recurring question of how the brain works, face the present state of the art in computational methods for analyzing huge amounts of data (Big Data) and the so-called conectome, arising from graph theoretical methods applied to small-world networks.

In chapter seven, Luiz A. Baccalá, with collaboration of Koichi Sameshima, reviews the concept of Granger causality, making clear how the associated "consistent prediction" improvement is different from usual definitions of causality, which require some physical linking mechanism or a criterion of intervention. The importance of such time series analysis is that it can uncover the directedness of multiple neural links underlying brain areas which show significant correlated activity.

João José Neto presents in chapter eight a thorough overview of the general computational strategies for simulating natural systems, with special consideration of the framework of adaptivity. This framework is implemented via rule-based devices of event-driven models, and adaptivity means that the set of rules can change over time in response to inputs. All the knowledge contained in the simulated phenomenon is encoded exclusively on the set of rules describing its operation. Learning thus amounts to the encoding of novel information into the database of rules. Adaptive procedures are therefore those that perform learning and forgetting.

Mariana Broens, João Moraes & Arturo Forner-Cordero discuss in chapter nine the internet of things, which includes the internet of robotic things. They characterize the phenomenon as a social affordance, describe the dissemination of this new social interaction, and explore some of its ethical implications.

Chapter ten, authored by Samuel Bellini-Leite, brings an overview of the frame problem in artificial intelligence, i.e. how an information processing system can deal in a relevant way with a changing environment. He lists twelve sub-instances of the problem, and argues that the "revisionist strategy" of taking these sub-instances seriously is the best path for cognitive science, a strategy that is consistent with a pluralist approach to the field, in which "all flowers may blossom".

Joao Kogler Jr. & Paulo Santos discuss, in chapter eleven, the geometrical and logical aspects of a cognitive agent, which distinguishes it from merely reactive and perceptive agents. What singles out the cognitive agent is the use of context-invariant information that breeds knowledge. In their approach, the perceptual system is formed by a chain of processes dealing with contextual information, beginning at the sensorial system and ending at the motor system. The perceptual chain is constrained not only at both extremities, but its transformations are geometrically connected with "holonomic" constraint propagation. On the other hand, the cognitive system comprises self-constrained processes that operate with learned transformations, so that the geometry of cognition is "non-holonomic". Shifting to the logical approach to the problem, "qualitative spatial reasoning" is unveiled in terms of "regional connection calculus", which describes how spatial regions are connected. The approach, as pointed by the authors, has not fared well with experimental data from cognitive psychology, which is expected since the formal approach is an abstraction from perception. Building on suggestions from advantaged properties of representations, a formalism is proposed which represents qualitative location by partitioning the visual scene around an agent into sectors, and providing relations among them, so that they form a language capable of expressing the spatial organization of the scene.

In chapter twelve, João Ranhel, Emilio Del Moral & Márcio Lobo discuss in detail a spiking neural network model in which the spikes propagate with delays and entail the formation of neural cell assemblies that fire in synchrony, functioning as coincidence detectors. They show how the model implements logical functions, memory, plasticity, and decision-making, working with central pattern generators.

André Leclerc & William Pickering (chapter thirteen) discuss two different views in philosophy of language. The first part investigates the linguistic explanation of "meaning", and argues that the traditional approach based on disquotational truth conditions is insufficient for natural languages. The consideration of the intentions of the speaker is also insufficient, so one must include in the analysis *intuitive* truth conditions, in which the agent grasps immediately the speaker's expectations and plans. The second part of the paper discusses how linguistics has been explored from the perspective of complex systems theory. The notion of an emergent property is essential, and such an approach has been combined with the usage-base theory of linguistics, although it is not inconsistent with generativism.

Ivo Ibri, João Queiroz & Vinicius Romanini (chapter fourteen) explore different philosophical issues related to the semiotic theory of Charles Peirce. The first point made is that pragmatism is an indispensable framework for semiotics. Next, the semiotic theory of mind is presented as a variety of situated and distributed cognition, as a middle position between representationalism and its negation. Finally, Peirce's concept of information is analyzed, showing how cognition may be construed in terms of potential and actual information, leading to a concept of cognition understood as a general and pervasive feature of the intelligible Universe.

Alfredo Pereira Jr., Jonas Coelho & Osvaldo Pessoa Jr. (chapter fifteen) present three different monist views on the mind-body problem. "Triple-aspect monism" considers that there are three main phases of actualization of the potentialities in Nature: the physical, the informational, and the conscious. The "double-face view" assumes that conscious mind and brain are irreducible to each other, stressing not only that the conscious mind is dependent on the brain, but that changes in the brain are also dependent on the conscious mind. "Qualitative physicalism" adopts the mind-brain identity thesis, and defends the view that subjective qualia are actual physical attributes of some region of the brain.

Peter Claessens & João Sato (chapter sixteen) review probabilistic modeling in machine learning, arguing that Bayesian statistical inference is not only a methodological tool for cognitive science, but might also be a theoretical description of the actual working of the brain. Still, Bayesian statistics should be complemented with other techniques. "Robot doctors" tend to be increasingly used, but the medical doctor should be able to understand how a decision was made by diagnostic support systems. The authors also point out the similarities between Deep Learning and the developmental stages of human beings.

Diego Zilio, Flávio Corrêa da Silva & Luciano Silva (chapter seventeen) explore different issues related to computational models of cognition. The first point made is that there is a significant difference between a computational model and an experimental model in biology and psychology: while the latter reproduce the mechanisms of the actual phenomenon being studied, the first only reproduces the behavior of the phenomenon. The paper also examines the distinction between deep artificial intelligence (AI), shallow (or biologically-inspired) AI, and mimicking AI, advocating an integration of the three approaches. As a case study, they present a probability calculus to be used in emotional user interfaces, based on algebraic semiotics.

In chapter eighteen, Loula, Lobo Netto Muñoz & Gudwin discuss fundamental questions concerning cognitive architectures, such as the definition of a "cognitive system", taken to be a system which aims at explicitly modeling human cognitive functions, including perception and action. Artificial intelligence that should not be called "cognitive" is also discussed. The importance of language for high-level performance of cognitive systems research is analyzed, and the frontiers of research in the field are explored.

Fábio Simões-de-Souza (chapter nineteen), in a short text that complements chapter twelve, summarizes the state of art of research on computer models of spike-timing-dependent plasticity, an important mechanism of long-term synaptic plasticity, essential for learning and memory.

Chapter twenty, by Rômulo Ferreira & Maria Inês Nogueira, closes the volume exploring issues in the history of neuroscience. In the first section of the paper, problems related to the formation and development of scientific concepts are examined, emphasizing how concepts such as "nervous cell" change with time. A distinction is drawn between concepts that have a clear referent from those that do not. In the second section, some methodological and philosophical issues of current research in neuroscience are examined.

Acknowledgements

First and foremost, we thank Alex Krueger for his kind invitation for publishing, and for having encouraged us to elaborate a book with the proceedings of the 10th Brazilian International Meeting of Cognitive Science (EBICC 2015). We very much appreciated the experience of working with Vernon Press, and we thank their editorial staff, for the tireless effort provided throughout the months of preparation of the book. We particularly acknowledge the careful supervision, support and patience of Argiris Legatos, for whom we owe much of the success. And we also thank the cover designer for the excellent artwork.

The achievement of this book would not be materialized without the effort of the several authors that provided their contributions in updated form for this publication, and we thank and congratulate them for the excellent work.

All the contents of this book originated as results of the meeting brought about by EBICC 2015, organized and promoted by the Brazilian Society of Cognitive Science (SBCC), with the partnership of the Polytechnic School of Engineering of the University of São Paulo. We thank the Director of Polytechnic School, Professor José Roberto Castilho Piqueira, for his commitment and personal effort in providing this partnership and support. We are largely indebted to several Departments of the University of São Paulo, for their joint effort and support in the local organization and production of the event, namely the Departments of Philosophy, of Electronic Systems Engineering, of Computer Engineering, and of Mechatronics. We thank Professors Antonio C. Seabra, Antonio M. Saraiva, José Reinaldo Silva, Ricardo L.A. Rocha, and Vinicius Romanini, for their personal efforts on providing local resources and facilities for the event, and also Professors Walter Teixeira Lima Jr., Marcos Antonio Alves, and Ítala Maria Loffredo D'Ottaviano, for helping out with the organization and financial support for some speakers.

We express our gratitude for the technical co-sponsorship of the Institute of Electrical and Electronics Engineers (IEEE) Computational Intelligence Society, provided by the South-Brazil Section Chapter, and we thank its President, Professor Emilio Del Moral Hernandez for his kind support. In addition, we thank Easy Chair (easychair.org) for hosting EBICC 2015's paper submission and evaluation services, Eventbrite (eventbrite.com) for hosting the subscription services, and Laboratory of Integrated Systems (LSI) of the Polytechnic School of the University of São Paulo, for hosting the event web site.

Finally, we want to acknowledge and thank for the financial support provided by FAPESP (Fundação de Amparo à Pesquisa do Estado de São Paulo), CAPES (Coordenação de Aperfeiçoamento de Pessoal de Nível Superior, of the Federal

Government of Brazil), NVIDIA Corporation, and CEST (Centro de Estudos Sociedade e Tecnologia da Universidade de São Paulo).

Many other people contributed for the success of the event, that lead to this book, and we apologize for not having acknowledged them nominally here, however we provided a full list of contributors at the EBICC 2015 web site (http://www.lsi.usp.br/ebicc-2015/).

Chapter 1

Autonomous action in complex mechanical systems: a real dilemma?

Maria Eunice Q. Gonzalez
Department of Philosophy
São Paulo State University (Unesp), Marília, SP, Brazil
gonzalezquilici@gmail.com

Abstract

In this paper we investigate the relationship between human *autonomous actions* and activities performed by complex mechanical informational systems such as self-organizing autonomous robots. Ethical problems related to the contemporary rapid development of these kinds of systems will be addressed, with special emphasis on the usage of autonomous robots in human social environments. In this context, the general problem that will guide the present analysis can be characterized as follows: Is there any (in)compatibility between the concepts of human *autonomous action* and autonomous *mechanical agency*? We argue that the *paradigm of complexity*, which unites interdisciplinary research perspectives in the areas of Philosophy, Physics, Biology, Robotics, Complex Systems Science, and Cognitive Science, amongst others, offers a useful conceptual framework for analysis of the problem in question.

> *The reason why it may be wise to distrust the political judgment of scientists qua scientists is not primarily their lack of "character" – that they did not refuse to create atomic weapons – or their naiveté – that they did not understand that once these weapons were developed they would be the last to be consulted about their use – but precisely the fact that they move in a world where speech has lost its power. And whatever men do know or experience can make sense only to the extent that it can be spoken about. [...] Closer at hand and perhaps equally decisive is another no less threatening event. This is the advent of automation, which in a few decades probably will empty the factories and*

> *liberate mankind from its oldest and natural burden, the burden of labouring and the bondage to necessity [...].*
> (Hannah Arendt, [1958] 1998, p. 4)

1. Introduction

Can a complex informational system, such as a robot, act autonomously? Answers to this question reveal strong polarization between dualists and materialists, which has been growing for centuries. Thus, a negative answer would be immediately given to this question in the 17th century, supported by theological and philosophical traditions. This would be the case, for example, of the (mind/body) dualist Cartesian view, which attributes to the soul the ability to move a (mechanical) body. Since robots do not have a soul, they could not, in principle, act, and, much less, act autonomously. According to this view, the movement of automata should necessarily be dependent on the command of their human inventors.

Assuming that only humans could act autonomously, in contrast to animals/machines that move around propelled by external forces impinging on their physical systems, René Descartes argued that there would be two tests able to always distinguish a human being from other animals or machines: the test of language and the test of action.

In part V of the *Discourse on the method*, Descartes argues that:

> *[...] if any such machines had the organs and outward shape of a monkey or of some other animal that doesn't have reason, we couldn't tell that they didn't possess entirely the same nature as these animals; whereas if any such machines bore a resemblance to our bodies and imitated as many of our actions as was practically possible, we would still have two very sure signs that they were nevertheless not real men. (1) The first is that they could never use words or other constructed signs, as we do to declare our thoughts to others. We can easily conceive of a machine so constructed that it utters words, and even utters words that correspond to bodily actions that will cause a change in its organs [...]; but not that such a machine should produce different sequences of words so as to give an appropriately meaningful answer to*

> *whatever is said in its presence – which is something that the dullest of men can do.* (Descartes, [1637] 2007, part V, p. 22)

With respect to the test of action, he claims that:

> *[...] even though such machines might do some things as well as we do them, or perhaps even better, they would be bound to fail in others; and that would show us that they weren't acting through understanding but only from the disposition of their organs. For whereas reason is a universal instrument that can be used in all kinds of situations, these organs need some particular disposition for each particular action; hence it is practically impossible for a machine to have enough different organs to make it act in all the contingencies of life in the way our reason makes us act.* (Descartes, [1637] 2007, part V, p. 22)

Descartes stresses that we should not confuse speech or actions with the natural movements of physical systems whose organization might imitate human actions or speech; he claims that parrots, for example, can repeat words, but they cannot speak as humans in such a way that they would understand what they say in different situations. In the same way, he envisioned the possibility of the creation of automata that could imitate human actions, in specific circumstances, but that they would necessarily lack the ability to act diversely and adequately in response to different situations in life, as humans generally can, with the help of reason.

In short, Descartes' claim of the radical limitations in speech and action inherent to machines is grounded on his (mind/body) dualist presupposition of their lack of soul and reason, which are supposedly responsible for the exclusively human abilities. However, this dualist presupposition has been strongly rejected by materialists of different types. In the 18th century, Julien de La Mettrie, for example, in *L'homme machine* (1747), developed several arguments to support his hypothesis that humans are not different from machines: not only do they act mechanically, but they are themselves machines! As a radical materialist, he claimed that human thoughts, language, and actions result from the complex organization of matter.

La Mettrie's ([1747] 1912) arguments in defence of a mechanistic approach to human action, language, and mind in general, are not the object of the present investigation, but it is worth mentioning that they have been supported by contemporary researchers in the areas of Artificial Intelligence, Cognitive Science,

Philosophy of Mind, and Robotics, who develop techniques to model intelligent activities related to action and speech in automated informational systems.

In the early 1950s, for example, in his article "Computing machinery and intelligence", Alan Turing considered the possibility of creating intelligent machines that, instead of storing a huge amount of information and following strict rules, could learn as children, by means of evolutionary processes of reinforcement and punishment, amongst others. Turing (1950, p. 460) argues for a mechanistic (functionalist) approach to intelligent behavior, claiming that:

> *[...] An important feature of a learning machine is that its teacher will often be very largely ignorant of quite what is going on inside, although he may still be able to some extent to predict his pupil's behaviour. This should apply most strongly to the later education of a machine arising from a child machine of well-tried design (or programme). This is in clear contrast with normal procedure when using a machine to do computations: one's object is then to have a clear mental picture of the state of the machine at each moment in the computation.*

In short, according to Turing's view, if a learning machine could execute a task in such a way that its results are not distinguishable from those obtained by an intelligent human, then this machine would have passed in the Turing Test and its behavior should be considered intelligent, independent of its internal functional structure.

Turing's notion of learning machines was, until the middle of the last century, interpreted as a rather polemical metaphor, but contemporarily it seems to have become a reality in the conjoined areas of Artificial Intelligence and Robotics. The rapid development of machines in these areas has introduced a new agenda of technical and philosophical problems related to the influence of learning machines in human society. In what follows, we are going to investigate, from a Complex Systems perspective, possible ethical consequences of the development of learning machines in terms of human autonomous individual and collective actions.

2. On autonomous learning machines: a view from the Complex Systems Paradigm

The Complex Systems paradigm is a recent interdisciplinary field of investigation that comprises at least two main branches (Zoya & Aguirre, 2011). The first mainly involves research in the areas of Philosophy and the Cognitive, Information, Social, and Environmental Sciences. The topics investigated in this branch focus on the influence of order, disorder, self-organization, and emergence in the structuring of individual/social actions, and their relation to the environment (Furtado et al., 2015; Cilliers & Preiser, 2010; Quilici-Gonzalez et al., 2014; Zoya & Aguirre, 2011). The second branch is mainly developed in the areas of Physics, Mathematics, Biology, Economics, Engineering, and Robotics, providing a framework for modelling aspects of adaptive multiple-scale systems such as social webs, collective intelligence, and weather forecasting, amongst others (Bourgine, 2013; Haken, 1988; Quilici-Gonzalez et al., 2010; Ladyman et al., 2012; Weaver, 1948).

Despite their different perspectives, both branches of the Complex Systems paradigm agree on the following minimalist characterization of complex systems:

> *A complex system is a group or organization which is made up of many interacting parts. [...] In such systems the individual parts – called components or agents – and the interactions between them often lead to large-scale behaviors that are not easily predicted from a knowledge of only the behavior of the individual agents. Such collective effects are called emergent behaviors.* (Mitchell & Newman, 2002, p. 190)

The dynamic interaction between the parts of a complex system distinguishes it from other types of systems: in complex systems, the interaction between the constituent elements creates a relation of dependence that is fundamental for the final organization of the system. In this circumstance, information of the behavior of isolated individual elements is not sufficient to understand the collective properties that emerge from the interconnections established in the communication amongst the elements.

One of the difficulties in predicting the behavior of complex systems is that they involve, in general, nonlinear feedback, which might allow the emergence of unfolding interconnected subsystems at multiple scales. From a formal perspective, nonlinearity can be described as follows:

> *A system is linear if one can add any two solutions to the equations that describe it and obtain another, and multiply any solution by any factor and obtain another. Nonlinearity means that this superposition principle does not apply. [...] Non-linearity in the equations of motion may have the*

> *consequence that small differences in the values of the initial conditions may make for radically different macrostates.* (Ladyman et al., 2012, pp. 4-5)

Multiple-scale dynamics is an inherent characteristic of complex systems that involves nonlinearity and, when this is the case, the *paradigm of complex systems*, with its interdisciplinary perspectives, offers a useful conceptual framework for its analysis. In this context, our working hypothesis is that the problem about the (in)compatibility between the concepts of human autonomous action and autonomous mechanical agency, could be better understood from the Complex Systems perspective.

Traditionally, philosophical studies on the nature of autonomy distinguish between (a) autonomy of the will, and (b) autonomy of action.

Autonomy of the will is in general characterized by philosophers in connection with free action, as the rational human capacity of making non-coerced, unselfish, and well-informed decisions. Kant (1785), for example, considered that autonomy of the will (associated with duty) is a central mark of human dignity; it is a main characteristic of our rational life, which would make no sense without it. Legal characterizations relate autonomy of the will to the liability of agents for the consequences of their autonomous actions.

Autonomous action, in turn, can be described as the human rational ability to establish independent, non-coerced, decisions, free of external imposed determinants, in such a way that her/his *intentional* conduct is (or could be) self-governed. This minimalist preliminary characterization of autonomous action, in terms of self-governing activities of a "super agent", is the object of several controversies. One central difficulty concerns the requisite of rationality, which is the object of debate among compatibilist and incompatibilist philosophers. As Buss (2012, p. 655) states:

> *Just as autonomous agency is compatible with stupidity and thoughtlessness, so too it is compatible with ambivalence, regret, disappointment, frustration, and self-criticism. Without any impediment to our capacity for self-government, we can determine that the best thing to do under the circumstances is something we sorely regret having to do.*

Another difficulty concerns the regress problem. As pointed out by Buss (2012, p. 656):

> *If I must do something in order to govern my own agency, then must I not also do something in order to govern this*

> *self-governing activity? If I must sanction a motive in order for it to be truly my own, then must I not also sanction this sanctioning attitude? Under what conditions, we want to know, is it truly the agent herself who determines how she exercises her agential capacities?*

In relation to the regress problem it is possible to argue that self-governing activities could result, in complex systems, from an emergent global self-regulating mechanism. From this perspective, as suggested by José Artur Quilici-Gonzalez (personal communication), the regress problem would be illusory, given that the sanction of a "sanctioning attitude" could result from the communicating dynamics of a self-organizational structure.

Despite the above controversies, the minimalist characterization of autonomous action is sufficient for discussion of the problem under analysis here. Likewise, even though (a) and (b) might be interconnected in human social life, or in dualist/monist debates, no necessary link is established between autonomy of the will and autonomy of action in the areas of Philosophy of Mind, Cognitive Science, and Robotics. Instead, degrees of autonomy are attributed to the action of a system due to its ability to act independent of an external control. This conception can be illustrated by the following passage from Murphy (2000, p. 4):

> *Autonomy means that a robot can adapt to changes in its environment on itself and continue to reach its goal [...] It can [...] operate under all reasonable conditions without recourse to an outside designer, operator or controller while handling unpredictable events in an environment or niche.*

The above view of the autonomy of robots, characterized in terms of their capacity for self-adaptation to unpredictable events, admits several degrees that might vary in proportion to their ability to deal with information available in the environment, independent of external control (Fong et al., 2003; Haselager, 2005; Murphy, 2000).

A robotic dog, from Boston Dynamics (Fish, 2011), exhibits some degree of autonomy in that it is able to climb hills, passing through steep and rugged terrain, and climb stairs independent of a human driver. Even when the robot gets out of balance and falls, it can self-stabilize and stand independently.

According to a video (see Stampler, 2015), this robot has the ability to self-stabilize without the help of its designers. It is one of the various types of mechanical models of complex informational systems under construction that exhibit degrees of autonomy (in the sense indicated by Murphy). This kind of system involves self-organization, circular causality, learning strategies, and decisions based on technical efficiency, which characterizes contemporary neo-mechanicism. But, in what sense is there self-organization in the actions of robots? What is meant by self-organization in this context?

Debrun (2009), following Ashby (1962), proposes two types of self-organization: primary and secondary. *Primary self-organization* happens whenever there is a process of spontaneous encounter between distinct elements, and their communicative interactions allow the emergence of a rough system or a basic type of organization. If learning and fine-tuning mechanisms operate in an already constituted system, then the *secondary self-organizing* process might be activated, and the primarily self-organized system will be spontaneously developed, independent of a central command or controller.

Both primary and secondary self-organization require interaction amongst the elements of a process, but secondary self-organization requires learning and adjusting mechanisms for restructuring a system or organization that already exists. In general terms:

> *There is self-organization every time that the appearance or the restructuring of a form, throughout a certain process, is due to the process itself – and to its intrinsic characteristics – and, only to a lesser degree, to its initial conditions, to the interchange with the environment, or to the causal presence of a supervising instance.* (Debrun, 2009, p. 34)

In the above sense, the robotic dog is not self-organized in the primary way because its system is completely structured by its designer, but it learns to adjust its movement as it moves around in the environment. Hence, the robotic dog exhibits the process of secondary self-organization.

Similar to the robotic dog, several types of adaptive self-organizing robots (ASORs) have been developed by the defense industries. Thus, for example, the Unmanned Combat Air Vehicle called Taranis is characterized as: "[...] a fully autonomous craft that can fly deep into enemy territory to collect intelligence, drop bombs and defend itself against manned and other unmanned enemy aircraft." (Cartwright, 2010).

ASORs and diverse types of artificial intelligent systems, such as controlled airplanes, trains, production lines, electricity power stations, water distribution systems, smart houses, and smart farms, amongst others, now control many aspects of human life (Quilici-Gonzalez et al., 2010, 2014). This increasing dependency between learning autonomous machines and human individual/social actions brings us back to the ethical and technical problems of autonomy: can this dependency severely restrict our autonomy? In which circumstances would it be possible to control the action of autonomous machines?

An answer to the above problems was given by Wiener (1960), who developed technical and ethical analyses on the topic of machine control, radically rejecting the layman idea that nothing can come out of the machine which has not been put into it. In agreement with Turing, he argues that learning machines can surpass the speed limitations of their designers, who could be unable to predict the future actions of machines after they have been submitted to a training process.

Wiener (1948) stresses that machines, analogously to living beings, can be programmed to adjust themselves to the environment and learn by means of past experience, logic, and communication strategies involving nonlinear feedback. According to him, the importance of nonlinear feedback, in both living organisms and machines, is that it is at the origins of the processes of both learning and reproduction:

> *Two of the phenomena which we consider to be characteristic of living systems are the power to learn and the power to reproduce themselves. These properties, different as they appear, are intimately related to one another. An animal that learns is one which is capable of being transformed by its past environment into a different being and is therefore adjustable to its environment within its individual lifetime. An animal that multiplies is able to create other animals in its own likeness at least approximately, although not so completely in its own likeness that they cannot vary in the course of time.* (Wiener, [1948] 1996, p. 160)

The possibility of uncontrollable self-reproduction in learning machines is one of the risks to human autonomous action that is pointed out by Wiener. Given the intricacy of the controlling aspects involved in autonomous robots, it seems that critical discussions on this topic should be on the agenda of interdisciplinary research projects related to ethical consequences of their proliferation.

3. Concluding remarks: Possible ethical consequences of the proliferation of autonomous robots

On July 25, 2016, *BBC Technology News* exhibited a robot developed by researchers of Stanford University, called Jackrabbot (Fig. 1.1), “which is learning how to be socially aware and move through pedestrian environments by observing behavior." Stanford University's Alexandre Alahi explained that, differently from robots that follow well-defined rules, Jackrabbot "can learn on the fly", learning "social conventions of the place it is in at the time" (BBC, 2016).

The relevant novelty brought about by this news is that robots like Jackrabbot are not just learning to move, but learning, with a certain degree of autonomy, how to move by observing embedded embodied human actions. This novelty involves not only physical, but also *communicating informational* strategies concerning human social habits.

According to our understanding, the assemblage of self-organization, nonlinear feedback, and cultural informational strategies in the development of learning machines radically distinguishes the Cartesian type of mechanicism from the contemporary neo-mechanicism initiated with Cybernetics (Wiener, 1948, 1960) and Artificial Intelligence (Turing, 1950).

Figure 1.1: Picture of Jackrabbot, from Stanford University's Computational Vision and Geometry Lab, a robot that exhibits a degree of autonomy by learning how to interact with people walking in the street (Hopping, 2016; permission for publishing granted by Silvio Savarese, SAIL - Toyota Center for AI Research, Stanford University).

In short, a central distinction between the 17th century kind of mechanicism, exemplified by the creation of clocks, mechanical ducks, and other fancy artefacts, and contemporary neo-mechanicism, is that the latter allows the manufacture of complex informational systems with several degrees of autonomy.

When learning strategies and decision procedures (based mainly on technical efficiency) are introduced in complex informational mechanical systems, significant consequences for human autonomous action might be on their way. But, is there a real (in)compatibility between the concepts of human autonomous action and autonomous mechanical agency?

From the robotics perspective, as indicated by Murphy (2000), there is no necessary incompatibility between the concepts of autonomous human action and self-organizing mechanical action, because both might be characterized in terms of degrees, and it is up to humans to establish how autonomous a machine could be, and when it should be turned off. However, as emphasized by Wiener ([1948] 1996, p. 176):

> *Again and again I have heard the statement that learning machines cannot subject us to any new dangers, because we can turn them off when we feel like it. But can we? To turn a machine off effectively, we must be in possession of information as to whether the danger point has come.*

A complicating factor concerning the problem of machine control is, according to Wiener (1960), directly related to time scales. Given that humans and machines operate on distinct time scales (humans are, in several ways, much slower than machines), the coordination between them might bring about unforeseen difficulties. Deeply concerned with moral problems of the development of autonomous machines, which operate with nonlinear feedback, he invites us to consider the coordination problem in the context of human actions in the long term:

> *[...] Problems of the same sort arise whenever two control operators on very different time scales act together, irrespective of which system is the faster and which system is the slower. This leaves us the much more directly moral question: what are the moral problems when man as an individual operates in connection with the controlled process of a much slower time scale, such as a portion of political history or – our main subject of inquiry – the development of science?* (Wiener, 1960, p. 1358)

This difficult moral question formulated by Wiener signals the importance of constant critical investigations on the history of possible human errors that have affected the development of societies. However, it seems (too?) early to decide in the case of the technical problem related to the human ability to deal with information necessary to control the development of learning autonomous machines. Thus, the technical problem of the possible (in)compatibility between the concepts of human autonomous action and self-organizing mechanical agency remains unsolved.

In relation to the ethical difficulties, there might be incompatibilities between human autonomous action and mechanical agency related to responsibility. Thus, for example, if an autonomous robot kills somebody, who is responsible? What kind of moral values should be imprinted in the machine that takes decisions in unusual moral situations? The dilemma we are faced with is that autonomous machines can be useful in many situations, assisting disabled people, helping in search and rescue missions, retrieving people in very difficult areas where unpredictable disasters occur, and so on. So, are we facing a real dilemma?

Our provisional answer to these queries is that human autonomy may be threatened by the proliferation of autonomous machines, depending on our ability to understand the dynamics of the technical, social, and political organizations in which we are immersed. According to the Complex Systems perspective sketched here, there are ways of analyzing the emergent properties of the multiple-scale interactions between humans and autonomous machines in order to anticipate likely future risks and consequences. As suggested by Bourgine (2013), "the aim [of complex systems research] is to act in order to produce the least possible uncertainty for responsible innovation."

In this article, we have suggested that the rapid development of autonomous machines in our contemporary society has introduced a new agenda of technical, social, political, and philosophical problems concerning the influence of learning autonomous machines on human autonomous action. Recognition of the incompleteness of human knowledge requires interdisciplinary dialogues, which might allow fertile interfaces without the requirement of the unifying totality of the diversity (Souza Santos, 2007). From this perspective, it could be the right time to examine Wiener's ([1948] 1996, p. 175) hypothesis that: "There is nothing more dangerous to contemplate than World War III. It is worth considering whether part of the danger may not be intrinsic in the unguarded use of learning machines."[1]

[1] Thanks to Guiou Kobayashi (UFABC), José Artur Quilici-Gonzalez (UFABC), Mariana Claudia Broens (UNESP), and Vivian Urquidi (USP), who helped with the development of the ideas presented in this paper; and to colleagues from GAEC (UNESP) and CLE (UNICAMP) for their inspiring discussions. This research was supported by the Brazilian agencies FAPESP and CNPq.

Bibliography

Arendt, H ([1958] 1998). *The human condition.* 2nd ed. Chicago: University of Chicago Press. Available at: http://sduk.us/afterwork/arendt_the_human_condition.pdf.

Ashby, W.R. (1962). Principles of the self-organizing system. In: von Foerster, H. & Zopf, Jr., G.W. (eds.). *Principles of self-organization.* Oxford: Pergamon, pp. 255-278.

BBC (2016). Jackrabbot: why this robot is watching how you move. *BBC Technology News,* July 25. Available at: http://www.bbc.com/news/technology-36836243.

Bourgine, P. (2013). Complex systems science: lifelong education by research. Unpublished manuscript distributed at the XXV Congresso de Iniciação Científica da Unesp, in Barra Bonita, SP, Brazil.

Buss, S. (2012). Autonomous action: self-determination in the passive mode. *Ethics,* 122(4), 647-691.

Cartwright, J. (2010). Rise of robots and the future of war. *The Guardian,* November 21. Available at: http://www.theguardian.com/technology/2010/nov/21/military-robots-autonomous-machines.

Cilliers, P. & Preiser, R. (eds.) (2010). *Complexity, difference and identity: an ethical perspective.* New York: Springer.

Debrun, M. (2009). *Brazilian national identity and self-organization.* Coleção CLE, vol. 53. Campinas: Centro de Lógica, Epistemologia e História da Ciência, Unicamp.

Descartes, R. ([1637] 2007). *Discourse on the method of rightly conducting one's reason and seeking truth in the sciences.* Transl. J. Bennett, 31 pp. Available at: http://www.earlymoderntexts.com/assets/pdfs/descartes1637.pdf

Fish, E. (2011). AlphaDog carries heavy objects, is easier to train than a real dog. *PC World,* October 3. Available at: http://www.pcworld.com/article/240978/alphadog_carries_heavy_objects_is_easier_to_train_than_a_real_dog.html

Fong, T.; Nourbakhsh, I. & Dautenhahn, K. (2003). A survey of socially interactive robots. *Robotics and Autonomous Systems,* 42 (3-4), 143-166. Available at: http://www.cs.cmu.edu/~illah/PAPERS/socialroboticssurvey.pdf.

Furtado, B.A.; Sakowski, P.A.M. & Tóvolli, M.H. (eds.) (2015). *Modeling complex systems for public policies.* Brasília: IPEA (Institute for Applied Economic Research). Available at: http://www.ipea.gov.br/portal/index.php?option=com_content&view=article&id=25858.

Haken, H. (1988). *Information and self-organization: a macroscopic approach to complex systems.* Berlin: Springer.

Haselager, W.F.G. (2005). Robotics, philosophy and the problem of autonomy. *Pragmatics & Cognition,* 13, 515-532.

Hopping, C. (2016). Jackrabbot is an autonomously navigating robot prototype. *Robot's Champion,* August 4. Available at: http://robotschampion.com/jackrabbot/

Kant, I. ([1785] 2005). *Groundwork for the metaphysics of morals.* Transl. T.K. Abbott. Revised edition by L. Denis. Orchard Park (NY): Broadview Press. Available at: http://www.justiceharvard.org/resources/immanuel-kant-groundwork-for-the-metaphysics-of-morals-1785.

Ladyman, J.; Lambert, J. & Wiesner, K. (2012). What is a complex system? *PhilSci-Archive*, 39 pp. Available at: philsc-archive.pitt.edu/9044/4/LLWultimate.pdf.

La Mettrie, J.O. de ([1747] 1912). *Man a machine.* Transl. G.C. Bussey. Chicago: Open Court.

Mitchell, M. & Newman, M. (2002). Complexity theory. In: Pagel, M. (ed.). *Encyclopedia of evolution.* Vol. 1. New York: Oxford University Press, 190-192. Available as "Complex systems theory and evolution" at: http://web.cecs.pdx.edu/~mm/EncycOfEvolution.pdf.

Murphy, R.R. (2000). *Introduction to AI robotics.* Cambridge (MA): MIT Press.

Quilici-Gonzalez, J.A.; Kobayashi, G.; Broens, M.C. & Gonzalez, M.E.Q. (2010). Ubiquitous computing: any ethical implications? *International Journal of Technoethics*, 1(3), 11-23.

Quilici-Gonzalez, J.A.; Broens, M.C.; Gonzalez, M.E.Q. & Kobayashi, G. (2014). Complexity and information technologies: an ethical inquiry into human autonomous action. *Scientiae Studia*, 12 (special issue), 161-179.

Souza Santos, B. (2007). Beyond abyssal thinking: from global lines to ecologies of knowledges. *Review*, 30(1), 45-89.

Stampler, L. (2015). Googles's robot dog brings us one step closer to the end of humanity. *Time*, February 10. Available at: http://time.com/3703243/google-boston-dynamics-robot-dog.

Turing, A. (1950). Computing machinery and intelligence. *Mind*, 59 (236), 433-460.

Weaver, W. (1948). Science and complexity. *American Scientist*, 36: 536-544.

Wiener, N. ([1948] 1996). *Cybernetics: or control and communication in the animal and machine.* 2nd ed. Cambridge (MA): MIT Press.

——— (1960). Some moral and technical consequences of automation. *Science*, 131 (3410), 1355-1358. Available at: http://www.nyu.edu/projects/nissenbaum/papers/Wiener.pdf.

Zoya, L.G.R. & Aguirre, J.L. (2011). Teorías de la complejidad y ciencias sociales: nuevas estrategias epistemológicas y metodológicas. *Nómadas: Revista Crítica de Ciencias Sociales y Juridicas*, 30, 1-20. Available at: http://pendientedemigracion.ucm.es/info/nomadas/30/rdzzoya_aguirre.pdf.

Chapter 2

A slim defense of narrow content

Frederick Adams
Department of Linguistics & Cognitive Science
University of Delaware, Newark, DE, U.S.A.
fa@udel.edu

Abstract

This paper considers Gabriel Segal's recent attacks on content externalism. Segal argues that the existence of empty kind terms refutes content externalism. He argues that an obvious rescue of saying that empty kind terms have no content might save the theory but for the fact that thoughts involving those empty kind terms are involved in perfectly meaningful behavior. Then Segal gives several examples of such terms and such meaningful behavior that they explain. Hence, he claims empty kind terms cannot be meaningless, and thus externalism is false. We show why Segal's examples and arguments still fail to establish that content externalism is false.

1. Slim support for narrow content

In his recent book, Gabriel Segal (2000) argues against externalists' accounts of cognitive content and for a kind of narrow, individualistic content. Segal defends a kind of content on which "being in a state with a specific cognitive content does not essentially involve standing in any real relation to anything external" (Segal, 2000, p. 11). And he defends a local supervenience thesis on which "if two beings are identical in respect of their microstructural properties, then they must be identical in respect of their cognitive contents" (Segal, 2000, p. 14). Contrary to many narrow content theorists, he holds that narrow content is a variety of ordinary representation. For example, in a Twin-Earth scenario[1] where Zoe's "diamond" concept was formed on diamonds and Twin-Zoe's "diamond" concept was formed on Twin-diamonds, Segal's (2000, p. 19) "view is that both Zoe's diamond concepts apply to both diamonds and twin diamonds". If true, in Segal's

[1] In such scenarios, these items are visually indistinguishable at a time prior to scientific tests to discriminate them.

example it seems to us that both Zoes have garden-variety broad, disjunctive content – not narrow cognitive content. There is nothing narrow about concepts of diamonds (or Twin-diamonds). So, we have a hard time accepting Segal's narrow content rather than what the tradition would call broad, disjunctive content.

We will not dwell on his positive account of narrow content here. This is in part because he claims not to have a theory of content to offer (Segal, 2000, p. 20), and in part because he says his "basic strategy is to undermine [...] the most popular and influential externalist theses and show that an internalist alternative is workable and attractive" (Segal, 2000, p. 20). We focus our attention on his arguments against externalism. We believe his main argument against externalism is easily met. If we are right that Segal's negative argument against externalism is easily met, then his strategy of arguing indirectly for narrow content by exposing difficulties with broad content turns out to offer slim support indeed for narrow content.

3. Segal's argument sketch

Empty terms and concepts are the test case for the dispute between externalists and internalists about content. Prior to Segal, Boghossian (1997) used his "Dry Earth" example to make a similar challenge to externalism – we discuss Boghossian's arguments below. And there have been other such attacks as well. Similarly, empty names have been viewed as the test case for direct reference theories of proper names for many years[2]. So, it is correct for Segal to look to empty kind terms and concepts as the front line in his attack on radical externalism.

Segal presents the externalists with a dilemma in accounting for the mere existence of empty kind terms and concepts. He asks what would happen to the meanings of kind terms and the concepts they express if the terms were vacuous. He asks us to imagine Twin-Earth scenarios where terms such as "water," "aluminum," "topaz," "quark," and "polio" are empty on Twin-Earth but not empty on Earth. Following Boghossian (1997) we can imagine that those on Twin-Earth do not know that the terms are empty (vacuous). They may be systematically deluded or deceived or just mistaken about whether such terms refer.

Segal maintains that the radical externalist must say either: (*a*) that such terms express no concept or (*b*) they express a concept "different form the one they

[2] See Everett (2003) and Braun (1993).

actually express" (Segal, 2000, p. 31).[3] We take the latter to be impossible, but what Segal may be saying is that they express a concept different from the one supposed by the user (thinker).

Here is a parallel with empty names.[4] We would take Segal to maintain that for the externalist, "Santa Claus" either (*a*) has no content or (*b*) if it does express a content, it expresses a content other than that supposed by the speaker. For example, "Santa lives at the North Pole" either expresses that "___ lives at the North Pole" or imparts but does not express that there is a jolly fat man who wears a red suit and brings presents on Christmas Eve and who lives at the North Pole. In the first case, "Santa Claus" semantically expresses *no content,* though there is content expressed by the remainder of the sentence. In the second case, we claim that the use of "Santa Claus" *pragmatically imparts* something, but *does not express semantically* what one would have expected, viz. that Santa Claus lives at the North Pole.[5] We will say more about this parallel as we go. For we maintain that it provides the key to understanding empty natural kind terms and concepts.

4. No content

Let's start with the "no content" horn of the supposed dilemma for externalists. Segal believes that the very existence of empty kind terms and concepts should be something of an embarrassment for externalists.[6] This we deny. We agree with Segal that empty kind terms exist. Many kinds of empty names exist. Kind terms are a kind of name (in so far as they are a type of linguistic entity). Semantically they are empty. Syntactically, they may play the role of a common noun. So, there are indeed linguistic entities that play the role of natural kind terms and are intended by their users to refer to natural kinds (without success).[7]

Why say, as Segal does, that maintaining that empty natural kind terms have no content is "courageous?" We suspect that it is because, for referentialists and externalists generally, there are going to be counter-intuitive results. Such contrary intuitions have to be explained away. We see this with direct reference theories and empty proper names no less than with externalism and empty natural kind terms. So, we maintain that, at bottom, Segal is exploiting the counterintuitiveness of externalism when he says "there are numerous empty

[3] We don't think these disjuncts are exclusive. Indeed, we think the radical externalist should embrace both (a) and (b) as will become clear in what follows.

[4] See our Adams & Dietrich (2004) for details.

[5] Indeed, we did not just express a complete proposition, but the form of words best *suggests* the proposition intended.

[6] He calls it "catastrophic" (Segal, 2000, p. 33).

[7] With the possible exception of fiction (Adams, Fuller & Stecker, 1997).

kind terms that we must take to express concepts [...] they are pervasive and significant features of human cognition" (Segal, 2000, p. 33).

Here is an example of a corresponding counterintuitive result from empty proper names and direct reference theories of them. Suppose one maintained that the very existence of empty names, such as "Vulcan," "Santa," "Sherlock Holmes," is an embarrassment for theories of direct reference.[8] On such theories, one may hold that "Vulcan does not exist" does not express a truth because "Vulcan" is an empty name. As such, it contributes no object to a proposition. So, when one employs a vacuous proper name in this negative existential sentence there is at most an incomplete proposition expressed of the form: there is not an x such that x = ___ (where one cannot fill the blank). Hence, there is no truth (or falsehood) expressed using the term "Vulcan." This surely is counterintuitive and has led philosophers to say that such cases spell doom for theories of direct reference. However, theories of direct reference may *explain away* the recalcitrant intuition. They may explain it away by pointing to the lore (character, stereotype) associated with the term Vulcan consisting of descriptions such as "the tenth planet," "the planet between Mercury and the sun," "a planet causing perturbations in Mercury's orbit". When one says "Vulcan does not exist" one literally expresses no truth, but one pragmatically imparts an associated proposition such as "that there is no tenth planet." The associated description, though not what is literally semantically expressed, is an imparted truth. It is the truth of the pragmatically imparted information that explains a hearer's (or speaker's) taking a truth to have been uttered.[9] Hence, one is able to explain away the counter-intuitiveness of the theory, and one is able to resist temptation to posit some extension-independent content for the term "Vulcan."

5. Content other than expressed

Returning to natural kind terms and the other horn of the dilemma, we suggest a parallel account that explains away the counter-intuitiveness of maintaining that empty kind terms semantically express no content. Lore (character, stereotype) associated with the kind terms will have content and may be pragmatically expressed or accessed when employing empty kind terms. Any cognitive content that appears to emanate from the empty kind term actually has its source in the associated terms and descriptions. We will now turn our attention to some of Segal's examples and explain how the "no content" (*a*, above, section 3) view works. At the same time, we will explain how this incorporates the view that a

---

[8] See Michael Devitt (1989).

[9] See Adams & Dietrich (2004) for details and references.

different kind of content other than what was expressed is the content that is imparted (*b*, above).

6. Examples

6.1 Bekong

At one point Segal introduces an empty kind term "Bekong." This term derives from a group known as the Fang (a people who live mainly in Gabon and Cameroon). The Fang believe Bekong to be a type of ghost entity from which people can directly receive inspiration or messages and who can cause people illness or misfortune. They believe the Bekong to be the ghosts of their ancestors.

Segal maintains that it is not an option to doubt that the Fang have a concept in the concept *Bekong* (even if empty). The concept plays an important role in the thoughts and behavior of the Fang.

We maintain that the Fang's term *Bekong* is itself empty (on the assumption that there are no ghosts in Gabon and Cameroon). And since the term is empty, there is literally no semantic content expressed in speech or thought involving this term. We suspect that Segal would ask "how can you deny that the Fang have a concept Bekong?" There is a syntactic structure in their heads and it is causing thoughts and behavior, so how can you deny that they have a concept?

Our reply is that they have a syntactic structure or name in thought and speech. This name is playing a syntactic role. But since it lacks a referent, it literally has no content itself. However, it does play a causal role in the speech and behavior of the Fang, and the role it plays is explained by the cognitive content of the associated descriptions and lore (character, stereotype). Since the associated descriptions have normal broad content, there is no difficulty explaining the role played by the empty Bekong concept in terms of the non-empty content of the associated concepts and meanings.

Segal's description of the ways the Fang think of Bekong is underdescribed, but we will provide some plausible elaboration. Here are some reasonable guesses. The Fang likely believe the Bekong "are the sources of the voices they hear in their dreams." They likely believe the Bekong "are the causes of the premature deaths of their infants." Perhaps they also believe the Bekong "are invisible causes of disease, plague, and famine." So, there are many descriptions that Fang associate with the term "Bekong."

On a referentialist/externalist account of the meanings of names or natural kind terms these associated descriptions (lore, character, stereotype) do not constitute the meaning of a term for the usual Kripkean (Kripke, 1972) reasons. But these descriptions can constitute the only contents associated with the empty terms. So, when a Fang utters a sentence such as "the Bekong are angry," the utterance literally expresses an incomplete proposition identified by the set consisting of a null subject and the property of being angry < ___, being angry>. The sentence is neither true nor false. And if an anthropologist studying the Bekong utters "Bekong don't exist," the anthropologist utters an incomplete proposition of the form that there is not an x such that x = ___ (where one cannot fill the blank).[10] The incomplete proposition is expressed using the term 'Bekong.' This utterance too is neither true nor false.

To explain the apparent truth of the negative existential sentence, we appeal to the information pragmatically imparted (but not literally expressed). Imparted is information to the effect that there are not invisible beings who speak directly to the Fang in their dreams. Or that there are not invisible conscious entities who cause disease, plague, and famine. And so on. Any apparent content expressed using the empty term "Bekong" is accounted for by appeal to the content of the associated descriptions (lore, character, stereotype).

This accounts for the mere existence of empty concepts on the radical externalist's view. We move now to the explanation of intentional behavior. Segal says "It is not a serious option to doubt that the Fang have an empty concept Bekong. It certainly appears that the concept plays an important role in their thought [...] beliefs, desires, and other cognitive states that motivate many of their activities" (Segal, 2000, p. 35) We agree that the Fang have an empty concept *Bekong.* We also agree that it plays an important role in the cognitive states and intentional activities of the Fang. What we deny, and what Segal has not defended, is that concepts that do not themselves literally express a content cannot play these important cognitive roles. On our view, the empty term "Bekong" becomes psychologically associated with meaningful descriptions (lore, character, stereotype). The cognitive role of the term "Bekong" is explained via (1) its cognitive association with the meaningful descriptions and (2) the literal broad semantic contents of those associated descriptions.[11]

Let's consider some examples. First, consider a case where a thought about the Bekong causes another thought. Suppose the Fang want to appease the Bekong and decide to best do so by giving them an offering. So, the fear of the Bekong causes the thought that to appease the Bekong it is good to give an offering. The first cognitive state may be the fear expressed by the English sentence "The Bekong are angry." This has the literal content identified with the ordered pair <____, being angry>. It is expressed using the vacuous term in thought "Bekong." How could this cause anything, if empty? It could by being psychologically

---

[10] If one maintains that all empty terms express the same content, we agree. Any differences are in pragmatically imparted, not semantically expressed, content (Adams & Dietrich, 2004).

[11] See Adams & Fuller (1992) and Adams et al. (1993).

associated with several other thoughts from the Bekong lore, such as the thought that there are invisible conscious beings who speak directly to us in our dreams and they are angry. It is associated with the thought that there are invisible conscious beings who cause premature deaths to our infants and those beings are angry. These associated thoughts cause other thoughts about how to avoid the bad affects of the anger of these beings. This leads to the intentions to appease these beings by giving an offering. Since the incomplete thought "the Bekong are angry" is cognitively linked to these other complete thoughts (even though they are false), it is quite easy to see how one thought (though empty) could cause other thoughts.

As for the explanation of the intentional activities, it should be quite easy to see how these incomplete thoughts about the Bekong can lead to complete thoughts and intentions about activities such as giving offerings (or other activities aimed at the non-existent Bekong). For these associated thoughts are complete thoughts (though false). So, we see no difficulty at all for the radical externalist who claims that empty concepts have no direct content. Any cognitive explanations of thought or behavior is accounted for in terms of the perfectly contentful lore associated with the empty natural kind terms. So, we do not disagree with Segal that there are empty natural kind concepts. We do disagree that their existence gives any good reason to believe in the existence of extension-independent (narrow) content. Our account above makes no use of such content and does what Segal says cannot be done – provide a viable alternative to his view (Segal, 2000, p. 38).

6.2. ME (myalgic encephalomyelitis)

In a second example, Segal gives a more traditional style Twin-Earthian argument for the existence of narrow content. Earth Peter has the concept of ME (by its other name, CFS, chronic fatigue syndrome). Twin-Peter has the same concept, but on Twin-Earth ME doesn't exist.[12] From the twins' epistemic perspectives, everything about symptoms of "ME" on Earth and Twin-Earth are identical (both are associated with dizziness, rashes, aches, sensitivity to light and sound, cold sores, swellings and forgetfulness). Thus, the argument is that Peter and Twin-Peter must share some content to explain the sameness of their thoughts and behavior. Since their concepts do not share extensions, there must be some extension-independent contents that their concepts share.

---

[12] In the example, we are to imagine that "ME on Earth" turns out to refer to a natural kind, but not so on Twin Earth. Segal says ME doesn't exist on Twin Earth either because there is no such disease or because "ME" refers to a motley of causes that do not comprise a natural kind.

Segal suggests that the assumption that Peter and Twin-Peter share the same concept of ME "would work perfectly well" (Segal, 2000, p. 41). Here he makes the standard assumption that the best explanation of the sameness of intentional behavior is that the twins share intentional content (in this case, the concept of ME). He is appealing to the idea that the best explanation of why the concepts of the twins have the same cognitive causal roles is that they share the same content. And "if the two concepts had only wide contents, then it is not at all obvious why their cognitive roles should be the same" (Segal, 2000, p. 43).

However, we *have an explanation* of why the roles would be the same even if their contents were different. The reason is that both twins associate the exact same lore with the terms "ME," even though on Earth the term is not empty and on Twin-Earth the term is empty. The content of the associated lore is identical and shared by the twins. The cognitive sameness of Earth Peter's ME-concept and Twin-Peter's ME-concept is to be explained in terms of the sameness of associated lore. The sameness of causal role in thought and behavior is to be explained in the same way.

So, suppose Peter thinks "ME is making me dizzy, forgetful, and causing a rash." This thought causes Peter to go to the doctor for medication. Peter's thought expresses the proposition identified by the n-tuple <Peter, causation, ME, dizziness, forgetfulness, and rash>. Peter associates with ME the descriptions "cause of dizziness, forgetfulness, and rash" (among other things). Peter existentially generalizes that he has a single kind of disease that is causing his symptoms and, therefore, he should seek a doctor. His thoughts are true and cause him to seek a doctor for help.

Twin Peter has a thought of the same form "ME is making me dizzy, forgetful, and causing a rash." Though he doesn't know it, his thought expresses the incomplete proposition identified by the n-tuple <Twin-Peter, causation, ___, dizziness, forgetfulness, and rash>. Twin-Peter's thought lacks a truth value. Since Twin-Peter does not know his thought is incomplete and lacks a truth value, he existentially generalizes that he has a single kind of disease that causes these symptoms and, therefore, he should seek a doctor. His thought that he has a single kind of disease is false yet causes him to seek a doctor for help. Twin-Peter has no disease ME, but he does have the same symptoms Peter associates with ME.[13] Twin-Peter's thoughts that he has these symptoms are not false. He, like Peter, believes he has a single disease called "ME" causing the symptoms. He like

[13] Please note that neither Peters' concept ME should be identified with their concepts of the symptoms of ME. The concept of the disease is not the concept of the symptoms. For example, the concept of measles is the concept of a virus that causes measles. It is not the concept of spots. SPOTS is the concept of spots. It is similar with ME. It is not the concept of its symptoms.

Peter goes to the doctor for help because he has this thought. Unlike Peter, his thought that he has ME is false. Like Peter, his thought that he has these symptoms is true.

Epistemically, since Twin-Peter is in exactly the same situation as Peter, his cognitive contents work the same in explaining the sameness of his (and Peter's) behavior.[14] The differences in the broad contents of their thoughts factor out precisely because the differences are epistemically opaque to the agents. The differences also factor out of the explanation of the sameness of their behavior.

Therefore, we see no difficulty whatsoever for the radical externalist to explain the sameness of behavior that there is between Peter and Twin-Peter. There is no need of extension-independent content to explain this sameness of the behavior of Twin-Earth and Earth twins. Though the content of their concepts is different (one being empty, one not), the contents of their associated concepts are the same. It is the contents of the associated concepts that share the same broad contents plus the cognitive association with the empty (and filled) concepts that explains the sameness of the twins' behavior (what sameness there is). Segal has the same strategy for explaining sameness of behavior of the twins as we do. It is just that he wants to find shared content in *the very concept of ME.* Whereas, we want to find only shared content in *the associated concepts and descriptions.* We see no reason to prefer his shared content to ours. Therefore, neither of these examples lends support to Segal's arguments against radical externalism of concepts.

We will return to two more of Segal's arguments when we consider the case of Swampman (and supervenience) below. Next, we wish to establish that our reply to Segal also satisfactorily handles Boghossian's "Dry Earth" example.

7. Dry Earth

In the context of arguing that there is an incompatibility between externalism and certain kinds of *a priori* knowledge of our thought contents, Boghossian (1997) introduces the example of Dry Earth. We won't here address the issue of compatibility with *a priori* knowledge.[15] Instead we will focus attention on Boghossian's claim that there must be a kind of content incompatible with externalism.

Boghossian introduced this as a case of an empty natural kind term and a powerful difficulty for externalism. Boghossian's (1997) Dry Earth is a Twin-Earth

---

[14] Of course, they go to different doctors because their broad thoughts are about different doctors.

[15] But we agree with McLaughlin & Tye (1998) that there is no incompatibility.

style example where Dry Earth is a "planet just like ours in which, although it very much seems to its inhabitants that there is a clear, tasteless and colorless liquid flowing in their rivers and taps to which they confidently take themselves to be applying the word 'water,' these appearances are systematically false and constitute a sort of pervasive collective mirage" (Boghossian, 1997, p. 170).

Boghossian (1997, p. 170) employs the example in order to raise the potentially embarrassing question "what concept, if any, should a Twin Earth externalist say would be expressed by tokens of the word 'water' on this Dry Earth?" Boghossian wants to say that on Dry Earth, 'water' expresses a concept even though it fails to refer. But what concept does it express? He suggests that one could say that it expresses a compound concept such as: "the clear, tasteless, colorless liquid that flows in the taps and the rivers around here now [...]" (Boghossian, 1997, p.171). Boghossian finds this suggestion to be plausible, but not available to the Twin Earth externalists because it would make 'water' an atomic concept when it referred, but a complex, non-atomic concept when it failed to refer (where the only difference was an external difference).[16]

Of course, we do not accept the compound description as a plausible candidate for the content of "water" on Dry Earth. So, we do not run afoul of the atomic/non-atomic problem. For us the concept is atomic but empty. Boghossian presses one who takes this line as follows: "What proposition—what truth condition—is expressed by sentences of the form, 'Water is wet', for example, as uttered on Dry Earth? What is it that gets said?" (Boghossian, 1997, p. 173)

Our answer, of course, is that "Water is wet" express the incomplete proposition identified with the order pair <___, being wet> (and expresses it using 'water'). This incomplete proposition is neither true nor false, but it is *what gets said*. Of course, we also maintain that on Dry Earth one who utters this sentence may pragmatically impart the false proposition that there is a cool colorless liquid that flows in the taps and rivers around here. This will be false, but will explain the apparent truth to Dry Earthlings and help explain their behavior (or the sameness of their behavior with Earthlings, such sameness as there is).

Boghossian further claims that such a line of response by the externalist is fatal to the claim that there is a concept in the first place, but this would be like saying that an empty proper name 'Vulcan' is not a name. True, it does not have a semantic value, but there is more to being a name than successfully naming. Names have syntactic form and can form psychological associations. Concepts do no less. Both do this whether they are filled or empty. So it is as much a mistake to claim that the externalist cannot admit the existence of empty concepts as to say that they cannot admit the existence of empty names. Thus, we reject

[16] We think this suggested content should not be attractive to Boghossian either because it is not extension-independent (narrow) content. It is merely conjunctive and broad. There is nothing non-externalistic about it.

Boghossian's claim that concepts must have non-empty extensions for externalists. Natural kind concepts can exist and be empty even on externalist accounts.

8. Swampman

We have answered Segal's and Boghossian's challenges to radical externalism. We will now conjoin two more arguments that Segal takes to support the existence of extension-independent content. Both are types of argument that appeal to supervenience and both are susceptible to the now standard swampman type of challenge – which we will present and defend.

Consider once again the "*ME*" example above. Let Peter and Twin-Peter be physical duplicates. According to the thesis of supervenience, mental properties supervene upon physical properties and there can't be a mental difference without a corresponding physical difference. There is no physical difference among the twins. So, Segal maintains, there is no difference in conceptual content. Peter's concept ME must be identical to Twin-Peter's concept ME (Segal, 2000, p. 44). Segal claims the physical neurological bases of the twins are identical and that the environmental differences of the twins are negligible. True, Peter's world includes the ME virus and Twin-Peter's doesn't. Instead it includes a motley of factors that produce Twin-Peter's symptoms. But Segal adds "[b]y assumption, his concept is empty. It has no extension. It does not apply to the motley of phenomena [...] Everything present on [Earth] that could be relevant is also present on [Twin-Earth]" (Segal, 2000, p. 45). So, Segal concludes that Peter and Twin-Peter share the same extension-independent concept ME.

Before we discuss this further, Segal's third argument adds to this only that for nativists (such as Chomskians), one may wish to consider diachronic conditions of concept acquisition. One would want to know how Peter and Twin-Peter acquired their concept of ME. "In order to explain this, we would have to develop a complex theory of [their] innate endowment and developmental history [...] how [they] learned about the diseases, and so on. But the key point is that whatever truth of the matter, everything specified by that theory would be present on [Earth and Twin-Earth]" (Segal, 2000, p. 46). So, whatever conditions are sufficient for Peter to acquire the concept of ME are available for Twin-Peter to acquire the concept. What this argument adds to the above is the innateness hypothesis. If concepts are innate, then they can develop in different kinds of environments, making it more plausible that the twins (in different environments) share the same concept. On the synchronic supervenience hypothesis, any physical difference in one environment could lead to a difference in mental content. On the diachronic, sameness of innate content may tolerate some differences in physical environments.

Our objection to these claims is that there is one crucial difference in the environments of the twins. Peter causally interacts with the ME virus. Twin-Peter does not. This will cause a difference in content. Peter's concept is *of ME.* Twin-Peter's concept is *of nothing* (though it is psychologically associated with concepts of causes and symptomatic properties associated with "ME").

Segal anticipates this reply and rejects it on the grounds that it is implausible because "developmental psychologists do not usually care whether the concepts they study are empty or not" (Segal, 2000, p. 47). We find this to be irrelevant since developmental psychologists do not usually discuss Twin-Earth style cases (nor do they usually care about a philosophical theory of semantic content).

The differences in content of the twins' concepts is not something that is epistemically accessible to the twins (or to developmental psychologists). So, there is no reason to suppose that the twins would notice the difference (or that developmental psychologists would – uninformed as they are about the differences on the twin planets). Since the concepts associated with ME are the same for the twins, the association of these same concepts can explain the sameness that exists for the twins' thoughts and behavior. That sameness would be sufficient for any developmental psychologist (or social scientist). However, the difference in causal relations to one's environment for the twins is crucial to the difference in content. One cannot think about a thing or property without interacting with the thing or property (or with one who has) – or so we maintain.

The swampman examples are the crucial test cases for the dispute we are now having with Segal. For these examples test our (and his) intuitions about just this issue: must one interact with a property or its instances to have thoughts the contents of which are that very property? Segal says no. We say yes.[17]

Along with twin cases, the test cases for theories of content have always included brains in vats or physical duplicates with no prior causal histories with an environment. Such is the example of Donald Davidson's (1987) swampman. Let Swamp-Peter be a physical duplicate of Peter (as is Twin-Peter), but with no prior causal history. Davidson has him fly together out of particles after lightning hits a stump in a swamp (hence, "Swampman"). If Segal's appeals to supervenience are correct, then Swamp-Peter should have the same concept ME as Peter (and Twin-Peter). Segal admits that a claim such as that Swamp-Peter has the concept ME is "less plausible than its analogue [for Twin-Peter] [...] because

[17] We should emphasize that we are mainly concerned with atomic concepts [...] concepts not built up out of other concepts. It may be possible to think about the property of being a unicorn without interacting with unicorns, but one must have the concept of horse and horn (or whatever).

[Swamp-Peter] has nothing resembling a history of acquisition of the concept" (Segal, 2000, p. 52).

Segal points out that if one assumes Swamp-Peter is subject to psychological explanations, then it is easier to maintain that Swamp-Peter should be said to share the same ME concept with Peter and Twin-Peter. However, we see no good reason to maintain initially – in the few hours after Swamp-Peter materializes – that his movements are subject to psychological explanation.[18] Just because there are syntactic structures in Swamp-Peter's head that are causing his bodily movements, that does not mean that his bodily movements constitute intentional behavior.[19]

Segal goes on to maintain that it would be "unprincipled" for the externalist to deny that Swamp-Peter has a concept of ME.[20] But his claim is based upon one's accepting his arguments that have gone before. Since we have systematically rejected those arguments, and provided principles for doing so, our denial here is hardly "unprincipled". If Swamp-Peter utters the sounds "ME hurts", we maintain that he has uttered nothing meaningful to him, nor has he expressed a proposition or a thought. This is not because he hasn't interacted with ME but because he has no intentional states whatsoever.

Segal's last gasp is the charge that to deny Swamp-Peter any thoughts or a concept of ME is "counterintuitive" (Segal, 2000, p. 55). If true, so be it. We are in the business of defending radical externalism. Our account of empty proper names and empty kind terms also requires that we hold negative existential sentences such as "Santa doesn't exist" or "phlogiston does not exist" to lack truth values (but be associated with imparted truths). We are willing to give up an intuition here or there to preserve a theory that us unified and treats all names (proper or common) the same. So, we are willing to embrace the counterintuitive result of maintaining that Swamp-Peter has no concept of ME.

---

[18] See Adams et al. (1990) and Adams & Aizawa (1997) for details. These papers have arguments that block Segal's defense of appeals to counterfactuals (Fodor-style) to explain Swamp-Peter's thought contents. We won't replay them here, since Segal (2000) relegates them to an afterthought section of Chapter 2.

[19] Compare Robot-Peter who has a non-intentional computer brain that is programmed to mimic the behavior of Peter. Robot-Peter would share bodily movements with Peter, but not intentional behavior. Swamp-Peter does the same. See also Dretske (1988) and Adams & Aizawa (1997).

[20] By the way, we are translating his "swamp Bedoin" example into his ME example, for simplicity of exposition.

9. Conclusion

In this paper, we defend the view that empty names and kind terms have no meaning because they have no bearers. Having no bearers, they cannot contribute an object or a property to the content of a proposition expressed on an occasion of use. Hence, they cannot be used to say or think things that are true or false.

We have given a mechanism to explain away any appearances that empty names or kind terms can express truths in such cases as negative existential sentences. We explain away the appearance of truth by virtue of the associated lore of empty names and the pragmatic implicatures generated by their use.

In addition, we have considered several objections. The objections of Segal and Boghossian focus not on empty proper names, but empty kind terms. Their arguments and examples challenge our version of externalism to explain not only what sentences employing empty terms express, but also how empty concepts can cause thoughts and intentional behavior. We explain how the contents of associated descriptions plus the syntactic objects that constitute tokens of empty terms in thought are up to these tasks.

Thus, we offer our radical externalist account of empty names and kind terms as an interesting and plausible contender in the semantics of names and kind terms. It is a unified view. On it, all names make the same kind of semantic contribution to thought or speech. Names of any kind contribute their bearers (objects or properties) to what is said or thought on an occasion of use (or thought). For atomic names and kind terms, empty terms have no meaning and contribute no object or property to a proposition expressed. However, associated lore can account for any apparent content or truth or falsity.[21]

[21] Many thanks to Anthony Everett, Eros Corazza, and Sarah Sawyer for comments on earlier versions of this paper.

Bibliography

Adams, F. & Aizawa, K. (1997). Rock beats scissors: historicalism fights back. *Analysis*, 57, 273-281.

Adams, F. & Dietrich, L. (2004). What's in a(n empty) name? *Pacific Philosophical Quarterly*, 85, 125-148.

Adams, F.; Drebushenko, D.; Fuller, G. & Stecker, R. (1990). Narrow content: Fodor's folly. *Mind & Language*, 5, 213-229.

Adams, F. & Fuller, G. (1992). Names, contents, and causes, *Mind & Language*, 7, 205-221.

Adams, F.; Fuller, G. & Stecker, R. (1993). Thoughts without objects. *Mind & Language*, 8, 90-104.

——— (1997). The semantics of fiction. *Pacific Philosophical Quarterly*, 78, 128-148.

Boghossian, P. (1997). What the externalist can know a priori. *Proceedings of the Aristotelian Society*, 97, 161-175.

Braun, D. (1993). Empty names. *Noûs*, 27, 449-469.

Davidson, D. (1987). Knowing one's own mind. *Proceedings and Addresses of the American Philosophical Association*, 60, 441-458.

Devitt, M. (1989). Against direct reference. *Midwest Studies in Philosophy*, 14, 206-240.

Dretske, F. (1988). *Explaining behavior*. Cambridge (MA): MIT/Bradford.

Evans, G. (1982). *The varieties of reference*. Oxford: Oxford University Press.

Everett, A. (2003). Empty names and 'gappy' propositions. *Philosophical Studies*, 116(1), 1-36.

Kripke, S. (1972). Naming and necessity. In: Davidson D. & Harman, G. (eds.). *Semantics and natural language*. Dordrecht: Reidel, pp. 253-355, 763-769.

McLaughlin, B. & Tye, M. (1998). Is content-externalism compatible with privileged access? *Philosophical Studies*, 107, 349-380.

Segal, G. (2000). *A slim book about narrow content*. Cambridge (MA): MIT/Bradford.

Chapter 3

Thinking with friends: embodied cognition and relational attention in friendship

Claus Emmeche
Department of Science Education, Faculty of Science
University of Copenhagen, Copenhagen, Denmark
cemmeche@ind.ku.dk

Abstract

Can research in situated and embodied cognition inform the study of interpersonal relations like friendship? And conversely, can friendship studies from disciplinary and interdisciplinary perspectives inspire research in cognitive science? These are the guiding questions for this chapter. Compared to cognition in animals, human cognition is situated in environments of cultural and societal systems, so the formation of a human self as a cognitive and emotional agent is a process involving distinct levels of embodied cognition, here called animate, anthropic and societal. Analyzing friendship as a social cognitive phenomenon, and bringing together observations and concepts from interdisciplinary studies of interpersonal relationships allows for a notion of *relational attention* to be developed. At least for some forms of friendships, the agents not merely attend to common interests; their perception is shared, mediated by the very relationship as an embodied activity of distributed cognition. An important example of this is collaboration in science and art, as when friends or colleagues work close together to solve problems or develop new forms of creative expression.

1. Introduction

Much research in cognitive science, including the recent advances in situated and embodied cognition, has focused upon cognitive processes involved in problem solving tasks of individual agents, and has said relatively little about how humans think together with other humans about problems related to themselves and the world. An exception could be the field called "social cognition" that is often seen as a sub-topic of social psychology, focusing on how people process, store, and apply information about other people and social situations; in other words, this field focuses upon the role that cognitive processes play in our social interactions.[1] Yet, this is still not about *thinking together*, in the sense of how human beings – through their dialogic exchanges in social, educational, scientific, political, artistic, or other work – build ideas, develop improved understanding of complex phenomena, and create new imaginations, concepts and theories, eventually mediated by relations of collegiality or friendship. This lack of focus is strange, given that cognitive science is a science, and science is par excellence an institution for thinking together under guidance of a set of methods or styles of research that define paradigmatic norms for research within a specific field.

This chapter presents material from an on-going interdisciplinary project[2] on friendship as a special social relation (both found in humans, in some non-human animals, and in the future perhaps also in situations involving social robots), in order to highlight some of the recurrent problems for a general cognitive science that aims at providing models of cognition that (1) cover different levels of complexity, (2) have high degrees of realism, and (3) have relevance for real-life situations involving human beings situated in specific cultures. That cognitive scientists can gather inspiration from unexpected sources, such as literary studies or sociology of science, should not surprise us, as a robust well-developed cognitive science must deal with human action and cognition in all its complexity, including the very social processes that define us as human beings and that are studied in their huge variety in the social sciences and the humanities (SSH). Before the SSH become just as integral parts of cognitive science as neuroscience and psychology, cognitive science cannot claim any completeness in its models of human cognition. Thus, friendship studies from the disciplinary and interdisciplinary perspectives of the SSH can inform research in cognitive science by elucidating some of the social and psychological complexity

[1] The cognitive processes investigated in social cognition are typically conceived of as taking place within the brains of individual beings, while only the origin of the information feeding into the system is recognized as 'social', distributed, and/or embodied. See, e.g., the entry in Wikipedia on social cognition (https://en.wikipedia.org/wiki/Social_cognition). Here it is said that the "major concerns of the approach are the processes involved in the perception, judgment, and memory of social stimuli; the effects of social and affective factors on information processing; and the behavioral and interpersonal consequences of cognitive processes."

[2] Initial ideas for the overall project were presented in 2013 at the VIII International Conference of Information, Knowledge and Action (EIICA), and the VII International Coloquium of Philosophy of Mind (CIFM), both in Marília, state of São Paulo, Brazil (cf. the abstract "Desafios da interdisciplinaridade no estudo da amizade" p. 25 in: *Caderno de Resumos do VIII Encontro Internacional de Informação, Conhecimento e Ação: "Informação e Complexidade: novos paradigmas no estudo do conhecimento e da ação"*, Marília, 2013). Parts of the projects are reported in Emmeche (2014, 2015a, 2015b, 2016). I thank the organizers of the EBICC 2015 conference for their help, hospitality and inspiration.

involved in social cognition among friends or close colleagues, and conversely, research in situated and embodied cognition can inspire the study of interpersonal relationships like friendship.

2. What is 'social' in friendship and social cognition?

Answering the question of what is the social aspect of embodied cognition demands distinguishing between different levels of embodied, embedded and extended cognition (Emmeche, 2007; Robbins & Aydede, 2009). At a rather basic level, *biological* embodiment enables self-organizing processes like metabolism and homeostasis, that are scaffolded by the genetic and epigenetic systems forming a plurality of biocodes. A more specific level is *animate* embodiment. Here, functional circles of motor- and perception-processes enables fine-tuned self-movement directed by animate intentionality, which is scaffolded by various neural representational systems. Such representational systems form an important biosemiotic precondition for those species of social animals that are living in groups and form stable bonds based on kinship and in some cases also friendship (Dag, 2011; Seyfarth & Cheney, 2012). At the animate level the 'social' in social cognition is often seen as the individual agent's capacity for having a 'theory of mind' or a proto-empathic 'mind-reading' capacity to infer and comprehend the intentions of other agents who are conspecifics or prey or predators.

A still more specific level only found in the human species is *anthropic* embodiment. Here, embodied cognition takes place in sociocultural groups of humans with culture-specific norm systems (like rules for family structures), enabling extensive collaboration, learning-based flexibility, and production of material as well as intangible cultural artefacts (like tools and narratives). These processes are all scaffolded by language as a particular system of communication and cognition that again catalyzes the emergence of *reflexivity* as a mediator between a personal self (as a part of human agency) and socio-cultural structure. The specific anthropological level of embodied cognition manifests itself in culture-specific ways of grasping and organizing the world in terms of myths, traditions, rites, and related procedures involved in reproducing the particular lifeforms of a given culture. The 'social' in social cognition at this level is a much more complex relational structure, tying the inner world of the agent's psyche to the outer sociocultural world of actions of the human group it belongs to. The emergence of language facilitates and augments new possibilities of thinking together: cognitive processes, that previously were tied to the actual being together of a few agents in a group of animals in a common situation, get externalized and generalized via the shared language system. This enables each agent to have a reflexivity about self and other, and roles and identities necessary for the attained social complexity. Embodied cognition in this anthropic level, scaffolded by the relational systems of language and culture, has traditionally been mediated within relatively close ethnic or tribal group-based relationships of family or kin, but it is also at this level that friendship in its genuine (yet culture-specific, cf. Hruschka, 2010) and full-blown form first appear.

Assuming the human species' history on Earth to be in the order of 100,000 years, it is only for a small recent section of this time span that it has organized itself by building extended civilizational systems. At this even more specific level, that we could call *societal* embodiment, we find the embedding of culture-specific ways of living a human life within larger institutional-organizational settings of functionally differentiated social systems (state, army, market, industry, a general public, civic organizations, institutional sectors of health, education, knowledge production, technology, communication systems), ultimately a global cosmopolitical sphere.[3] Here the 'social' in social cognition takes on new meanings and differentiates into many forms of sociality with different implications for social cognition. 'Thinking together' gets socialized and institutionalized in different ways (science is a great example, state governance is another), and a feature at this level of embodiment seems to be a more clear distinction between what is private and what is public in the affairs of a cognitive agent. Another feature is the still increasing capacities for distributing cognitive tasks to many agents interacting in large institutional systems. The emergent distinction between public and private interpersonal relationships tend to make friendship be considered, at least in modernity, as something private and informal, while for instance collegiality and marriage are (in addition to their informal aspects) also public relationships that are formally scaffolded by legal systems and rules.

Implied in this conception of different levels of embodied, extended and embedded cognition is a recognition of the limited validity of any single model cognitive science can offer when dealing with multiple levels of organization, especially when it comes to investigating anthropic and societal forms of semiotic

---

[3] The distinction between anthropic and societal forms of embodied cognition is made for analytical clarity, and is based upon the evolution and cultural history of humans. As it is hardly possible today to find human groups living in complete independence of larger institutional state-based systems, real-life examples of human embodied cognition will typically be societal ones, and these have (as their emergence-base) the anthropic forms, just like the emergence-base of anthropic embodied cognition is the animate form. Anthropic embodied cognition can thus be seen as an ideal type category that is rarely found 'pure', perhaps except for a few isolated Amerindian tribes in Amazonas. In this kind of categorization every 'higher' level includes in its material being the lower levels, meaning that anthropic embodied cognition is found everywhere in human life, only being overlaid by the societal form; just like being human includes the facts of also being an animal and being alive.

scaffoldings,[4] that is, how 'the social' in social cognition is supported by artificial tools for externalizing cognitive processes into social norms, rules, procedures, habits, rituals, etc., or simply supported by persons who are well-known friends of the agent. For instance, sociologist Margaret Archer found evidence for the existence of different modes of reflexivity (understood as an agent's continuing dialogue with herself and/or others nearby) characterizing an agent's personality. Especially those agents she called 'communicative reflexives' rely upon a form of distributed cognition, especially when they have to make the big decisions about what kinds of education, family form and work to pursue in their life (Archer, 2012; Emmeche, 2015a).

Friendship is hard to define but has been a topic of philosophical inquiry since classical antiquity (Konstan, 1997; Pakaluk, 1991), though for the long medieval and early modern periods it was not really a focus for philosophy.[5] Many have followed Aristotle in noting that although you may have many friends for 'pleasure' or 'use', like your business friends, you can have only a few genuine or close friends. This later notion of real friendship designates a social relationship between persons who mutually like to be together and share not only interests, experiences and actions, but also have some affection for each other, and care about the other person for his or her own sake. Such a notion is often taken to have a normative core meaning,[6] implying an ideal of an almost faultless relationship of which the real world may provide more or often less perfect exemplars (compare other normative concepts, as that of a state, where real states are less than perfect and we even know 'failed states'). The specific content of that notion, the 'goodness' that a good friend exemplifies, is worth considering for social cognitive science. The limits to the numbers of close friends one may possibly have is plausibly related both to the limited cognitive capacity of an agent (apart from that agent's character or 'goodness') and to the social structure in which that agent is placed (e.g., being a slave or a prisoner physically limits

---

[4] For an introduction to the notion of semiotic scaffolding, see the special issue of *Biosemiotics*, 8(2), introduced by Hoffmeyer (2015).

[5] At least friendship was not in these periods considered as very interesting for moral theory; according to Mitias (2012), due to the prevailing intellectual paradigms at those times (Mitias' repetitive *tour de force* is more about the significance of such 'paradigms' for moral theory than about friendship). In philosophy as a western intellectual tradition, friendship has predominantly been considered from an ethical perspective (cf. 'virtue friendship' in Aristotle), but also other philosophical perspectives are of course possible. Recent philosophers have questioned the idea that the best examples of friendship are always founded in morality. Nehemas (2016) argues that criminals can be friends no less than the most virtuous humans.

[6] However, one should beware that any 'core' meaning of friendship as a type of relationship should be scrutinized for its historically variable context that affects the meaning of the term (cf. Caine, 2009; Österberg, 2010).

your friendship capacities). Thus, caring for another person for his or her own sake[7] is a form of love that, as other forms like parental love, is constrained both by the agent's economy of *relational attention* as an embodied cognitive and emotional process – that is, the available time to afford and the devoted interest to offer – and by the very ability of the agent to enter into adequate interactions (shared action, communication, and empathetic understanding) with the friend.

3. Seeing together – friendship as 'sunaisthetic'

Relational attention can be understood as having three aspects, one of which is historically specific to modernity; the other two are more universal. The first aspect refers to the specific form of attention realized in modern friendship that is often said to be characterized not just by the friends' mutual trust and affection, but also by their self-disclosure, their willingness to reveal 'private' aspects of the inner or intimate life to the friend, partaking in and sharing their human fragility and vulnerability.[8] This is in line with recent attempts by social theorists to characterize intimacy in modernity as 'pure relationships' (Giddens, 1992), attempts that have been criticized for a tendency to idealize such relationships (Jamieson, 1998), or to tie them too closely to a contemporary 'love ideology' and 'therapy culture' (Blatterer, 2015, p. 113ff).

The second, and more reflexive aspect of relational attention, compatible with the anthropic level of embodied cognition, is that friends not only enjoy one another's goodwill, trust, and affection; friends certainly share these things, but, to quote two anthropologists, "crucially – each not only knows this of the relationship, but also knows that the other knows it; this suite of interconnected

---

[7] Many scholars follow Aristotle in seeing this as an important feature of a genuine friendship, but not all agree with him that also the lesser types (friends for use or for pleasure) should be considered as friendship. Persons called friends only because they are useful or give you pleasure are not friends, according to Nehemas (2016). Thus, "instrumental friendship" is "a contradiction in terms" (ibid., p. 113) even though "no friendship is completely *un*instrumental" (p. 112). But a genuine friendship can grow slowly and real friendship comes in degrees (ibid., pp. 105, 109, 132, 137, 196). Thus, Nehemas combines a modern emphatic notion of friendship with a realism concerning friendships' varying intensities.

[8] Friendship scholars generally agree that (1) there are some specific characteristics to friendship in modern societies, in contrast to traditional ones; (2) the mentioned features (like mutual self-disclosure) are central here; and (3) this is tied to a sharper distinction in modernity between the private and the public in the life of the agents (e.g., Blatterer, 2015; Caine, 2009; Jusdanis, 2014, pp. 15ff, 78; Österberg, 2010). However, this agreement does not concern the explanations of this pattern, which may be overdetermined by several factors (cf. Konstan, 1997, pp. 11ff; Beer & Gardner, 2015).

characteristics means that friends can count on or trust one another, which thereby becomes relevant to their actions and strategies" (Beer & Gardner, 2015, p. 426). This basic feature of mutual knowledge about the relationship opens for an account of friendship that underlines friends' willingness to be interpreted or drawn by each other, and to include a friend's interpretation of oneself in one's self-understanding: "I see myself through you. We are thus, to some significant extent, each other's creators" (Cocking & Kennett, 1998, p. 509). In friendship, we are not only willing to engage in such a creative mutual interpretation, we are also willing to be directed by the other (ibid.), to go and watch a ballet for instance, not because I had a previous interest in it, but because you suggested we could see it together. Nehemas (2016, p. 223) follows this account when he claims our own friends have an idea of who we are or should be; so, what we see of ourselves in them is "a picture, an interpretation of who we are to them that we can use – not uncritically – to guide our self-formation".

By pointing to such shared activity of two friends experiencing their world, Flakne (2005) gives a phenomenological analysis of the genesis of a self-relation mediated through the friend, an analysis emphasizing the agents' embodied and embedded existence. She argues that it comes about by joint-perception, *sunaisthesis* (Aristotle's term), as the act of not only perceiving an external action or situation, but also "simultaneously perceiving the friend" – the third aspect of relational attention. Sunaisthesis is a codetermination in a reciprocal occurrence by and of two organisms, taking place through an act of double intentionality: both agents are oriented to a common object, but each is also oriented to the friend's perception of that object. I perceive my friend's perception of the object through his bodily manifestations in gesture, conduct, and, of course, speech. This social and pre-discursive aspect of perception is connected with the fact that "whatever we take pleasure in, we enjoy *more* when we are in the company of friends" (ibid., p. 51); thus, Flakne talks about the exuberant pleasure of sunaisthesis, in contrast to the merely reactive pleasure in perception. Also, when seen from this aspect of relational attention, it is evident why an agent cannot have many real friends (or extend the sunaisthetic experience to as many as possible), because an "active community of perception [*sunaistheeseoos*] must necessarily be in a smaller circle."[9]

Thus, the relational attention in genuine friendships involves not only dialogues with mutual self-disclosure, as a model of modern friendship proclaims, or a mutual reflexive knowledge of the very relationship as a foundation of trust, but

---

[9] Flakne (2005, p. 58); she quotes here Aristotle's *Eudemian Ethics,* 1245b22-24. Flakne's innovative analysis is not just an instance of Aristotle scholarship dealing with his philosophy of friendship; its major points on joint-perception deserve to be taken up by empirical cognitive science.

also this very basic awareness of being and seeing together that common perception can bring. Joint-attention as the third aspect of relational attention seems to be a basic part of social cognition, bridging the animate and the anthropic level of embodied cognition.

It can be argued that both the second and third aspect of relational attention as a social phenomenon underlies what Archer (2012, pp. 87-124) describes as 'the relational solution' of two people who are trying to reconfigure their individual concerns and priorities, in order to adapt them to each other for a life together, acknowledging the emergent 'relational goods' generated by their dovetailing their projects into a common whole. This kind of seeing and thinking together in such relationships can be seen as an instance of what Bennett (2010), in an account challenging individualistic conceptions of personhood, calls *plural agency*. To be a plural agent, the friends "must together hold a joint (albeit possibly implicit) understanding of the well-being of that plural agent as such and so of the joint cares and concerns that constitute it as a plural agent" (ibid., p. 281). Archer's and Bennett's accounts indicate how both friendship and joint cognition can be given a real ontological status on a supra-individual level.

Plural agency and joint-perception do not seem to require any particular kind of friendship or that the friends be very similar. 'Seeing together' can be cultivated in many relational contexts, like when colleagues collaborate over extended periods and discuss a project, problems, or a common vision. This reminds us of a study by the literary scholar Victor Luftig (1993), investigating collaborations between authors, poets, and artists, one of which was the friendship between the author Virginia Woolf and the painter and critic Roger Fry. Luftig (p. 209) notes that in Woolf's writings, as testified by her biography on Fry, 'seeing together' stands as the very characteristic activity of their relationship. In their discussions of art and art criticism, 'seeing pictures' took on an essential significance for them, for as Woolf remarked, "to have another pair of eyes to see with, another brain to argue with, was a very necessary process in making up [Fry's] mind" (ibid., p. 212). Luftig underscores that 'seeing' to them meant complement, not assimilation; their shared vision embraced conflict, it was not founded upon unanimity.

4. Thinking together – patience and empathy

Also, the recent study of inventive pairs by the writer Joshua Wolf Shenk points to the role of tensions in creative processes. Shenk (2014, p. 21) reminds us that "making new, beautiful, useful things is as much about discord as it is about union". Shenk's investigation, bounty with examples of creativity at work, provides important questions for social cognitive science about the nature of self, other, and distributed cognition in pairs. For instance, distributed cognition could in principle rely solely on internet communication to achieve virtual presence between the partners, but real collaboration seems to demand at least some amount of really being together of the two workmates. As an example of the importance of time spent together in scientific collaborations, Shenk mentions Daniel Kahneman, a Nobel-prize winning psychologist, and his work with Amos Tversky, a cognitive and mathematical psychologist. Their early work focused on the psychology of prediction and probability judgment, and later they worked together to explain irrational human economic choices, contributions that became a part of founding the field of behavioral economics. The two researchers ended up working at different universities, and though they spent, thereafter, every second week together, and "talked on phone multiple times a day – sometimes for hours – the collaboration eventually languished" (Shenk, 2014, p. 132). Kahneman later told that they had failed to appreciate how critically their successful work "depended on being together at the birth of every significant idea, on our rejection of any formal division of labour, and on the infinite patience that became a luxury when we could only meet periodically" (ibid.).[10]

Patience is generally considered to be the capacity to accept or tolerate delay, trouble, or suffering without getting angry or upset. When such subjective suffering has causes rooted not in one's own body or in nature external to it, but in interactions with another person, the cognitive effort of 'putting yourself in the other person's shoes' – or mental paradigm – typically helps to increase patience to the benefit of collaboration. This is not an argument for general patience in science! Again, empathetic understanding does not equal unisonous acceptance or lack of criticism, but is a mode of cognitive interaction in which a critical dialogue rests upon the capacity to understand, so far as possible, a dialogic partner's arguments, even though they may be considered false, misplaced or missing the point – simply to better be able to point out specific failures of reasoning. Science should not, of course, have patience with false beliefs or theories, but should directly expose their failures and prevent fallacies in its own reasoning. Cognitive science, however, must include an aim to develop models for the origins of fallacies, and also models explaining the role of empathy in distributed cognition, disregarding whether specific cognitive acts in fact contribute to reach false or true conclusions, and disregarding whether or not one

[10] That collaboration demands patience is testified by another of Kahneman's collaborations (cf. Emmeche, 2015b), i.e., the one with Gary Klein. In chapter 22 of Kahneman's 2011 bestseller on decision making – *Thinking, fast and slow* – he relates how he and Klein "had many discussions, resolved many disagreements, almost blew up more than once, wrote many drafts, became friends, and eventually published a joint article with a title that tells the story: 'Conditions for intuitive expertise: A failure to disagree'. Indeed, we did not encounter real issues on which we disagreed – but we did not really agree." Here, friendship and what Kahneman calls 'adversarial collaboration' went hand in hand.

of the partners in a research collaboration can see that the other's argument can only be true in a fiction.

Both friendship and literature nudge the self to engage in the world of another agent – or many other agents – by way of empathy, the ability to understand and share the feelings and thoughts of another and thus to feel compassion for that other. Humanities scholar Gregory Jusdanis (2014) explores such a similarity between fiction and friendship in his study of what he sees as a link between our need for friends and our engagement with art and literature. Both friendship and fiction demands a readiness to enter invented worlds, reading the mind of another; in both, "two uniquely human functions come into play, the faculty of imagination and the emotion of empathy" (ibid., p. 6). Jusdanis maintains that readers' ability "to project themselves into another situation or into the mind of another person is what links their engagement with fiction with their capacity for friendship. Both practices involve conjecture and analogical thinking – the search for resemblances" (ibid., pp. 29 ff).

5. Concluding remarks

To use the distinction of Robbins & Aydede (2009) between the embodiment thesis (cognition depends not just on the brain but also on the body), the embedding thesis (cognitive activity routinely exploits structure in the natural and social environment), and the extension thesis (the boundaries of cognition extend beyond the boundaries of individual), this chapter has indicated (1) the necessity also to distinguish between animate, anthropic and societal levels of embedding cognitive processes in a cultural and societal environment; (2) that interdisciplinary studies of human friendship and collaboration can inspire research in cognitive science; and (3) that a notion of *relational attention* in its different aspects can be a focus point for case studies of how human beings may 'think together'.

Bibliography

Archer, M.S. (2012). *The reflexive imperative in Late Modernity.* Cambridge (GB): Cambridge University Press.

Beer, B. & Gardner, D. (2015). Friendship, Anthropology of. In: Wright, J.D. (ed.). *International Encyclopedia of the Social and Behavioral Sciences.* 2nd ed. Amsterdam: Elsevier. Vol. 9, pp. 425-431.

Blatterer, H. (2015). *Everyday friendships: intimacy as freedom in a complex world.* Houndmills (GB): Palgrave Macmillan.

Bennett, B.W. (2010). *Love, friendship and the self: intimacy, identification, and the social nature of persons.* Oxford: Oxford University Press.

Dag, A.I. (2011). *Animal friendships.* Cambridge (GB): Cambridge University Press.

Caine, B. (ed.) (2009). *Friendship: a history.* London: Equinox.

Cocking, D. & Kennett, J. (1998). Friendship and the self. *Ethics,* 108(3), 502-527.

Emmeche, C. (2007). A biosemiotic note on organisms, animals, machines, cyborgs, and the quasi-autonomy of robots. *Pragmatics & Cognition,* 15(3), 455-483.

——— (2014). Robotic friendship: can a robot be a friend? *International Journal of Signs and Semiotic Systems,* 3(2), 26-42.

——— (2015a). Semiotic scaffolding of the social self in reflexivity and friendship. *Biosemiotics,* 8(2), 275-289.

——— (2015b). The borderology of friendship in academia. *Amity: The Journal of Friendship Studies,* 3(1), 40-59. Portuguese translation: (2015). O estudo das fronteiras da amizade na academia. In: Broens, M.C.; Moraes, J.A. & Souza, E.A. (eds.). *Informação, complexidade e auto-organização: estudos interdisciplinares.* Coleção CLE, vol. 73. Campinas: Centro de Lógica, Epistemologia e História da Ciência, Unicamp, pp. 145-176.

——— (2016). Friendship, love, and the borderology of interdisciplinarity. In: Emmeche, C.; Budtz Pedersen, D. & Stjernfelt, F. (eds.). *Mapping frontier research in the humanities.* London & New York: Bloomsbury Academic.

Flakne, A. (2005). Embodied and embedded: friendship and the sunaisthetic self. *Epoché,* 10(1), 37-63.

Giddens, A. (1992). *The transformation of intimacy. sexuality, love and eroticism in modern societies.* Cambridge (GB): Polity.

Hoffmeyer, J. (2015). Introduction: semiotic scaffolding. *Biosemiotics,* 8(2), 153-158.

Hruschka, D.J. (2010). *Friendship: development, ecology, and evolution of a relationship.* Berkeley: University of California Press.

Jamieson, L. (1998): *Intimac: personal relationships in modern societies.* Maldon (MA): Polity.

Jusdanis, G. (2014). *A tremendous thing: friendship from the* Iliad *to the Internet.* Ithaca (NY): Cornell University Press.

Kahneman, D. (2011). *Thinking, fast and slow.* New York: Farrar,Straus & Giroux.

Konstan, D. (1997). *Friendship in the classical world.* New York: Cambridge University Press.

Luftig, V. (1993). *Seeing together: friendship between the sexes in English writing, from Mill to Woolf.* Stanford: Stanford University Press.

Mitias, M.H. (2012). *Friendship: a central moral value.* Amsterdam: Rodopi.

Nehemas, A. (2016). *On friendship.* New York: Basic Books.

Österberg, E. (2010). *Friendship and love, ethics and politics: studies in the Medieval and Early Modern History.* Budapest: Central European University Press.

Pakaluk, M. (ed.) (1991). *Other selves: philosophers on friendship.* Indianapolis: Hackett.

Robbins, P. & Aydede, M. (2009). A short primer on situated cognition. In: Robbins, P. & Aydede, M. (eds.). *The Cambridge handbook of situated cognition.* Cambridge (GB): Cambridge University Press, pp. 3-10.

Seyfarth, R.M. & Cheney, D.L. (2012). The evolutionary origins of friendship, *Annual Review of Psychology*, 63, 153-177.

Shenk, J.W. (2014). *Powers of two: the essence of innovation in creative pairs.* London: John Murray.

Chapter 4

Neutral monism's scientism and its pragmatic limits

Sofia Inês Albornoz Stein
Graduate Program in Philosophy, Humanities School
Unisinos University, São Leopoldo, RS, Brazil
Researcher of CNPq
siastein@me.com

Abstract

In this paper, I will argue that neutral monism, understood from a pragmatic perspective, is an acceptable philosophical view – even if still a partly metaphysical one – which gives a good alternative interpretation of how perception and other mental processes occur. In order to prove my point, I will interpret Russell's neutral monism approaching it from a semantic and pragmatic perspective that relativizes the discourse of mental events and processes, making it plausible to think of mental events as something knowable and describable from different perspectives, without sustaining an idealistic standpoint. I begin by showing that the reference of psychological or physical languages could be the same when it comes to describing "raw feels" (see Feigl, [1958] 1967), even if the perspective varies. On my view, Bertrand Russell's neutral monism (1921) intends exactly this: to show how we can establish an explanation that leads from a third person point of view of the body and the brain to the first person perspective that "feels" or senses what science can only see as a physical process. I will then show why Russell's physicalism, interpreted as a pragmatic criticism of introspection and of first-person discourse on sense-data, can (a) clarify the kind of scientism Russell presupposed in his philosophy of mind (section 2), and (b) how this was linked to his metaphysical claims about sensations (section 3). Finally, I will discuss why one should not accept Chalmers' (1995) view that, if an informational state has two distinct aspects, namely conscious experience and the physical process, the most reasonable intermediate position between reductionism and substance dualism should be a property dualism. In any case, I will defend first that the physical-mental dichotomy is a linguistic one, obviously based on some evidence related to the differences between first- and third-person perspectives of "mental" processes, and second that to be able to progress in this area of research, in the area that tries to explain the human mind, we shouldn't presuppose that our current vocabulary about it can determine its substance and establish the exact limitations of our knowledge of the mind.

1. What is at stake?

Today we face a widespread dilemma: either we accept ordinary descriptions made by humans of their own subjective experiences (their "talk about impressions") and correlate them to what we can observe scientifically regarding the physical changes that they present, as if they were parallel events, or we try to investigate from a behaviorist point of view what happens physically and in terms of behavior when humans are in a certain perceptual position, and ignore the first-person description of mental states, linking initial physical behavior to final physical behavior. In any case, the link between subjective experiences and physical behavior (initial or final) is yet to be discovered and explained. Bertrand Russell (1921), when sustaining neutral monism, thought he had found this link in particular neutral elements that would synthetize mental sensations and physical sense data. Nowadays, the neurosciences can't prove the existence of sense data, but retain the empiricist goal of finding the causal network that goes from physical objects to their "mental" perception. This network has to include multiple forms of "representations" of properties of the physical objects. The observation of neuronal activation during the process of perception of physical objects led to the conclusion that perception goes through multiple stages and that, in a broad sense, objects are actually built gradually in our brain, through the many inter-relationships established between different parts of it, each one performing a different function. I will here discuss the possibility of interpreting Russell's neutral monism – so as to be able to retain some of its content – from a semantic and pragmatic perspective that relativizes the discourse of mental events and processes, making it plausible to think of mental events as something knowable and describable from different perspectives, without sustaining an idealistic standpoint.

As stated by Willard Quine (1966, p. 667), Russell in *Analysis of mind* (1921) and in *Analysis of matter* (1927) detaches himself from a strongly phenomenalistic form of logical atomism, which presupposes a subjective acquaintance with atomic sense-data as the foundation of all knowledge, and achieves a still imperfect naturalism, which is not yet a strong physicalism, but which interprets sense-data as neutral particulars and no longer as subjective atomic qualities. Quine claims that Russell drifted in the direction of naturalism because his description of the neutral elements of experience was dependent on natural science and on scientific findings, and these neutral elements were no longer objects of introspection.

According to Herbert Feigl ([1958] 1967, p. 59):

> *The distinctive mark of neutral monism is a conception of the "given" which (1) is subjectless, i.e., it does not allow for the use of the personal pronoun "I"; and (2) is "neutral" in the sense that the given is characterizable as neither "mental" nor "physical". It maintains that both mentalistic concepts (the concepts of psychology) and physical concepts (those of physics) are logically constituted out of the more basic concepts designating neutral data. Psychology and physics are here understood as more or less systematic knowledge both on the level of common life, and on the more advanced level of science.*

I won't reconstruct Feigl's position with regard to the mental or his interpretation of Russell's neutral monism here, but the above description interests me insofar as it (a) emphasizes the 'subjectless' aspect of sensations according to Russell's neutral monism, which helps to approximate the object of direct experience to the natural domain of the sciences, and (b) shows the pragmatic aspect of the discussion developed by the logical empiricists, which emphasizes the language, concepts, and laws used by the different sciences, which can be brought closer if their concepts are brought closer. Therefore, the logical empiricist standpoint is a critical standpoint, in Kant's sense: concepts and their properties are part of human knowledge and don't capture any essential traits of the world. Even so, a fundamental methodological problem remains: in any other natural science, there is intersubjective evidence that can ground agreement about how to apply a concept. In psychology and any other science or investigation of the so called "mental" processes, the content of the concepts or the evidence to which the concepts apply – in Kant's vocabulary: the intuitions –, even if "subjectless", seem, at first glance, not to be open to investigation from a third-person standpoint (a scientific standpoint). Therefore, sensations are usually called "mental", because of the apparent first-person privilege of their access.

Feigl explains what is today known as the explanatory gap between first- and third-person perspectives on sensations:

> *If a brain physiologist were equipped with the knowledge and devices that may be available a thousand years hence, and could investigate my brain processes and describe them in full detail, then he could formulate his findings in a*

> *neurophysiological language, and might even be able to produce a complete microphysical account in terms of atomic and subatomic concepts. In our logical analysis of the meanings of the word "physical" we have argued that the physical sciences consist of knowledge-claims-by-description. That is to say that the objects (targets, referents) of such knowledge claims are "triangulated" on the basis of various areas of observational (sensory) evidence. What these objects are acquaintancewise is left completely open as long as we remain within the frame of physical concept formation and theory construction. But, since in point of empirical fact, I am directly acquainted with the qualia of my own immediate experience, I happen to know (by acquaintance) what the neurophysiologist refers to when he talks about certain configurational aspects of my cerebral processes.* (Feigl, [1958] 1967, p. 83)

Feigl's logical empiricist standpoint, even if not behavioristic, is an ontological monist view, according to which there can be more than one perspective on sensations; but what the neurophysiologist investigates is the same object which the subject experiences by acquaintance. So, Feigl's view isn't a neutral monism – there is a subject and there is acquaintance experienced by her – but it is certainly a monism. In contrast to phenomenological views according to which introspective reports are irreducible to intersubjective scientific reports of brain activity because these reports talk about different "objects" (or different "properties" of a same substance), Feigl states:

> *Now, direct acquaintance with "private" raw feels is describable also in the intersubjective language of science. Its ultimate explanation may again have to refer to various cerebral areas, one of which (speaking for ease of exposition again dualistically) "corresponds" to sensing, another to judging, and possibly another yet corresponds to (introspective) reporting. I conclude that acquaintance statements differ only in the type and domain of evidence, but not in regard to their reference, from certain neurophysiological statements.* (Feigl, [1958] 1967, p. 95)

So, the *reference* of psychological or physical languages could be the same when it comes to describing "raw feels", even if the perspective varies. Russell (1921) is in agreement with this, which is what I shall argue in this paper.

2. Brain, mind and a pragmatic method

Historically, the introduction of sense-data had, according to Barry Maund (2003), two major motivations, one epistemological and one metaphysical. According to Maund, Russell's motivation was epistemological, because he was searching for a "logically indubitable foundation for knowledge-claims" (Maund, 2003, p. 75), and this could apparently be provided by "logically privileged epistemological access" to sense-data. The metaphysical motivation – not Russell's – would postulate sense-data as having ontological existence, but without providing an indubitable foundation for knowledge. It is nevertheless difficult to say if, when in 1921 Russell adheres to neutral monism, his motivation is strictly epistemological or ontological. I suppose that he had both motivations, because he wanted to explain our access to the physical world through our sensations and, at the same time, to establish the ontological status of these sensations, even if not necessarily in a metaphysical sense. Sciences also have their ontologies.

Bertrand Russell's neutral monism (1921) tries to show how we can establish an explanation that leads from a third person point of view of the body and the brain to the first person perspective that "feels" or senses what science can only see as a physical process. So, Russell is trying to answer an old question, already made by Descartes, for example, about how we can explain the relation between brain and consciousness. Today one would ask how it happens that conscious experiences emerge from neural processes, or supervenes upon them.

Russell (1921) isn't defending an idealism of sensations or a property dualism (cf. David Chalmers), which sustains the irreducibility of sensations to neuronal process. Russell's physicalism, interpreted as a pragmatic criticism of introspection and of first-person discourse on sense-data, can clarify the kind of scientism Russell presupposed in his philosophy of mind, and how this was linked to his metaphysical claims about sensations. This would, I think, be the best interpretation of Russell's neutral monism. This in turn would help to prevent the emergence of a strong dualist position in contemporary philosophy of mind. I will also discuss why one should not accept Chalmers' (1995) view that the most reasonable intermediate position between reductionism and substance dualism should be a property dualism, if an informational state has two distinct aspects, namely conscious experience and the physical process.

I am presupposing that there is no final evidence of a specific causal chain that leads from objects to sensory impressions and our rational capacities. Therefore, I agree with Donald Davidson (2001) that we don't yet have psychophysical laws that could link our rational capacities – or our sensitive experiences – to physicochemical events in the brain.

Why continue to worry about Russell's empiricist philosophy of mind? Because, his empiricist tenets and arguments, which include scientific and metaphysical claims as part of a coherent whole, seem, from one point of view, to rescue the philosophical classical tradition; and because of his beautiful and sensible arguments, regarding what were then new discoveries about the brain. But the background for our use of Russell's neutral monism will be a pragmatic view of knowledge, already present in Vienna's logical empiricism and, specifically, in Wilfrid Sellars' criticism of the Myth of the Given ([1956] 1997).

Not only consciousness or private subjective qualitative experiences are mysterious from a scientific point of view; any kind of scientific object is, to some degree, mysterious – because of our ways of accessing reality. Any access, be it through perception, through perception and language, through introspection, or simply feeling, with or without the use of propositional descriptions, is limited. Any naturalistic explanation of knowledge has to conclude this knowledge is established in a limited human way. Therefore, there is not just one puzzle for science and philosophy, there are endless puzzles.

3. Russell's neutral monism and science

According to Robert Tully (1988), Russell's conceptual development from 1914 to 1927 did not involve very radical conceptual changes. In 1914, Russell considers William James and Ernst Mach's neutral monism as a throwback to idealism, because the neutral elements that would link the material world to the experienced world would be sensations. However, Russell gradually became aware that the neutrality of these elements was exactly their capacity to be described by a variety of scientific laws, which could be called psychological or physical. According to Russell (1921), when he accepted neutral monism, he saw a relativity in scientific laws and how they are labeled; that is, in the future we can imagine new kinds of laws that aren't in a strict sense mental or physical. Today, neurosciences do not prove the *existence* of Russell's neutral elements, the *sensations*, which would at the same time be sensory data and sense experience. But today's neurosciences have plenty of data as evidence for a link between the impact of objects on our sensory neural surfaces – nerve endings – and our "subjective" experience of these encounters. Russell's neutral monism has also received attention in recent years because, contrary to Quine, for example, it doesn't adhere to a behavioristic approach to the mind, which would deny the need for "internal mental processes" (Chalmers, [1995] 2002, p. 92) to explain behavior.

From Joseph Levine's (1981) concept of an "explanatory gap" between physical and psychological processes, Chalmers infers the impossibility of doing the same kind of investigation of both. As such, Chalmers sustains in his writings in the 1990's that:

> *(A) Without assuming that sensations are also physical [explainable by Physics], the causal link between objects and private subjective perception would be jeopardized.*

Instead of seeing Russell's neutral monism as confirming this assumption, we must bear in mind that, according to Russell, even if (A) is true, (B) and (C) are also true:

> *(B) Sensations aren't determined, in their nature and content, by an association with an intentional and conscious subject.*

> *(C) Despite the differences between the laws of the physical and of the so-called psychological, the momentary individuals, or elements, in Mach's ([1886] 1959) vocabulary, need not be called strictly physical nor psychological. They would be 'neutral' and could be even explained by laws not yet known.*

Observing neural activation during processes of perception, of feelings, or of conceptualizing, neuroscientists conclude that these processes have various stages and part of these stages happen in the brain, in certain parts of it, in certain specific anatomical regions, but also that these regions get synaptically connected during the whole process.

Images of nerve cells in the brain don't show what is happening from a first-person point of view or what the experience of these neurological processes might be like. So perspective is, obviously, important. But this doesn't mean that there is a mental property that can't be investigated by empirical science. It just means that there are possible perspectives of something we call material or physical. Obviously access to one's own perceptions and sensations is still first-person only, which shows that Russell's statement (1921) that sensations are the intersection between the mental (object of the first-person perspective) and the physical (object of the third-person perspective) is still relevant.

According to Tully (1988), it was not neutral monism that led Russell to choose the expression "sensation" to replace "sense-datum"; apparently he was searching for a proper term to describe something not strictly speaking mental, and which could be described from the point of view of a comprehensive science of the human mind. Russell's decreasing emphasis on introspection, or on first-person privileged access to one's own experiences, and also on the concept of acquaintance in his writings prior to *The analysis of mind* (1921), such as in "On the nature of acquaintance" (1914a), "The relation of sense-data to physics" (1914b) and "On propositions: what they are and how they mean" (1919), contributed to his choice of neutral monism. Russell claims in "On propositions" that James' neutral monism made more sense, as he was beginning to realize the importance of linking the physical objects of scientific research, particularly the brain, to the investigation of mental activities.

Russell's distinction between perception and illusion shows that his explanation of sensory experiences cannot be an exclusive metaphysical idealist one:

> *When a mental occurrence can be regarded as an appearance of an object external to the brain, however irregular, or even as a confused appearance of several such objects, then we may regard it as having for its stimulus the object or objects in question, or their appearances at the sense-organ concerned. When [...] a mental occurrence has not sufficient connection with objects external to the brain to be regarded as an appearance of such objects, then its physical causation (if any) will have to be sought in the brain. In the former case it can be called a perception; in the latter it cannot be so called.* (Russell, 1921, p. 136)

So, there is a need to speak about the cause – i.e., the objects – of perceptual states so as to be able to compare this causal chain with another proximal causal chain that doesn't include a distal external object as the cause of the resulting experience. This isn't an idealist statement.

4. Pragmatics and ontology

According to Sellars ([1956] 1997), sense-data or sensations aren't, strictly speaking, objects, neither for a metaphysics of the mind nor for common sense. Sense-data are postulations, as Quine would say, of an empiricist philosophy. In addition to this, Sellars argues that even our ordinary speech about our own mental processes doesn't really consist in a strong realistic theory about these processes. Even if we admit, like Sellars, that "talk about impressions" could emerge from ordinary discourse, these impressions wouldn't necessarily be real objects that we could investigate in the context of science, especially in the context of the neurophysiological and physical sciences. One of the main reasons for Sellars to sustain such skepticism about reductive physicalism with regard to sense-data would be his pragmatic belief that different languages fit different purposes, even if sense-data could be postulated by theorists interested in explaining a "micro-theory of sentient organisms" and could, eventually, correspond to the concept of impressions of ordinary language:

> *[...] some philosophers have thought it obvious that we can expect that in the development of science it will become reasonable to identify all the concepts of behavior theory with definable terms in neurophysiological theory, and these, in turn, with definable terms in theoretical physics. It is important to realize that the second step of this prediction, at least, is either a truism or a mistake. [...]*
>
> *To ask how impressions fit together with electromagnetic fields, for example, is to ask a mistaken question. It is to mix the framework of molar behavior theory with the framework of the micro-theory of physical objects. The proper question is, rather, 'What would correspond in a micro-theory of sentient organisms to molar concepts pertaining to impressions?' And it is, I believe, in answer to this question that one would come upon the particulars which sense-datum theorists profess to find (by analysis) in the common-sense universe of discourse (cf. Section 23).* (Sellars, [1956] 1997, p. 113-114)

So, Sellars does not exclude the possibility of a description of sense-data which would be part of a scientific theory about perception. "Sense-data" could be part of technical vocabulary used to describe how sentient organisms react internally to external forces. However, this does not mean that the sensory data provide a specific reference for expressions such as "red" or "triangle". At the same time, sensation of redness or of a triangular form may be part of a micro-theory, which can help explain scientifically – with all the pragmatic limitations of a scientific investigation – the internal reaction of sentient organisms to the environment. Knowing what really exists and how it behaves, indeed, even knowing the limitations of our investigations and explanations, is important for every scientist, independently of her purposes from a practical point of view. This means that our descriptions and explanations can continue to approach what exists.

Chalmers (1995) claims that nobody knows why physical processes in the brain are concomitant to conscious experiences. Why is it, asks Chalmers, that when our brains process light of a certain wavelength we experience red? This isn't to say that Chalmers denies that conscious experience has its origins in the brain, but we don't have a clue, he says, why there is a conscious experience associated with certain brain functions:

> *Consciousness is a feature of the world over and above the physical features of the world.*
>
> *This is not to say it is a separate 'substance': the issue of what it would take to constitute a dualism of substances seems quite unclear to me. All we know is that there are properties of individuals in this world – the phenomenal properties – that are ontologically independent of physical properties.*
>
> *[...] the property dualism that I advocate involves fundamentally new features of the world. Because these properties are not even logically supervenient on microphysical properties, they are nonphysical in a much stronger sense. When I speak of property dualism and nonphysical properties, it is this stronger view and the stronger sense of nonphysicality that I have in mind.* (Chalmers, 1996, p. 125)

But, how is it that Chalmers can state so categorically that what he calls phenomenal properties are not physical and not even a substance, but are, nevertheless, "ontologically independent"? Isn't this assertion unclear? I won't pursue Chalmers' metaphysical thought experiments to defend his property dualism here, mainly because I am worried about scientific evidence for his separation of the physical and the non-physical and of substances and properties. For me, these distinctions, in the context of his philosophy of mind, are unclear. I will clarify why they are unclear: If Chalmers states that consciousness is a feature of the world, "over and above the physical features", which feature is it? That is, how can there be a feature without its being a feature *of* something? What kind of feature can exist in isolation of any substance? It is such a strange feature that it is very difficult to accept the possibility of its existence. In our world, and in the manner we are able, within the evolutionary process of our species, to describe our world, any feature has to be a feature of something. Therefore, we are left with just two possibilities: (*i*) consciousness is something apart from our world, and therefore not investigable with our usual skills, which turns it into a metaphysical subject; or (*ii*) consciousness is part of our world, but remains something yet to be explained by science. Maybe Chalmers has chosen the first possibility and is trying to defend it through thought experiments that sound rationally acceptable, as in many metaphysical arguments. But even if this is so, he must base his argumentation on private individual qualitative experiences, and, thus far, we know that these are part of our world, of the world of empirical scientific evidences. Why should it be different?

In any case, the physical-mental dichotomy is a linguistic one, obviously based on some evidence related to the differences between first- and third-person perspectives of "mental" processes. But to be able to progress in this area of research, in the area that tries to explain the human mind, we shouldn't presuppose that our current vocabulary about it can determine its substance and establish the exact limitations of our knowledge of the mind.

5. Final Remarks

If we argue, as we have done, in favor of a kind of ontological monism in the philosophy of mind that is close to science, and also in favor of a kind of pragmatic view of knowledge, we are presupposing, obviously, that knowledge of objects, of our own body, of our own interactions with the world, and of other bodies and their interactions, results in self-knowledge. This includes knowledge of our own mind, even if this knowledge has to be mediated by language, theories, scientific instruments and methods. Denying science possible access to our private "life" and to its intrinsic nature is no different from denying science understanding of phenomena that aren't familiar to us. But human science has shown that even phenomena that seemed very strange and mysterious at the beginning can start to make sense gradually to our limited way of knowing. Maybe some "mysteries" will remain mysteries, but then we should, from the start, be open about this. If we don't know what something is, we can't say it is phenomenal or a feature or a substance or non-physical. We simply can't say anything about it. And, as Gareth Evans states, we know a lot about our "phenomenal" processes, processes that emerge from physical processes:

> *In the first place, we have what might be described as a general capacity to perceive our own bodies, although this can be broken down into several distinguishable capacities: our proprioceptive sense, our sense of balance, of heat and*

> *cold, and of pressure. Each of these modes of perception appears to give rise to judgements which are immune to error through misidentification.* (Evans, 1982, p. 220)

Phenomenal processes result from physical processes. And, I would say, they are part of these physical processes, because without what we call the phenomenal, or mental, or psychological part of the process, the process as a whole wouldn't achieve its purposes. Common sense presupposes that phenomenal states are needed to link other physical processes. For example, fear can cause an appropriate reaction, maybe in the form of reasoning followed by an action. Therefore, phenomenal sensations and feelings seem to be needed to begin an action. Entertaining thoughts and imagining are perhaps only possible because of the phenomenal experiences we have. These experiences seem to establish links between different parts of our bodies, and permit retaining and identifying different stimuli from different sources. Are phenomenal experiences dispensable? Could we not feel anything and still be what we are? I suppose conscious experiences are necessary to maintain our complex way of integrating our complex body to a complex environment. But they are no more than that.

A second way of gaining knowledge of our own physical properties is:

> *[...] the way in which we are able to know our position, orientation, and relation to other objects in the world upon the basis of our perceptions of the world. Included here are such things as: knowing that one is in one's own bedroom by perceiving and recognizing the room and its contents; knowing that there is a tree in front of one, or to the right or left, by seeing it; and so on.* (Evans, 1982, p. 222)

Evans' considerations "tell against the common idea that our conception of ourselves 'from the first-person perspective' is a conception of a thinking, feeling, and perceiving thing, and not necessarily of a physical thing located in space" (Evans, 1982, p. 222). So, again, knowledge of our own mind is not achieved through introspection or through rational self-reflection alone. Knowledge about our own mind is mediated by proprioception, perception, language, and theories. Pragmatics helps to keep in view our need for language and theories. Recent theories of embodied cognition remind us of the importance of the body and of our perception of it so as to begin any reasoning from our own "mental" experiences.

To sum up:

(1) Knowledge about my body is basic self-knowledge.

(2) Perceptual content and the resulting memories (images, representations, etc.) contain information about our bodily interactions with the world.

(3) Reports of mental events and processes are derivative of the words and sentences of intersubjective learning that is simultaneous to the memorizing of perceptual content (the question of the necessity of a common-sense theory of mental events, impressions, and thoughts for that to happen remains an open question).

(4) Reports of thoughts and intentions, for example, are derivative of acts of thoughts and intentions about perceptual content.

(5) If we consider that even self-reports need intersubjective language-learning and theories, it is easy to infer that scientific third-person reports aren't so diverse from self-reports. Pragmatics has clarified this.

Even if there remains a great number of open questions about the role of phenomenal experiences in our physical interactions with the world, the great amount of evidence in favor of its utility to basic processes as giving attention to something specific in a scene, memorizing, recalling, and reasoning, seems to point to a scientific way of investigating consciousness and phenomenal mental processes that denies consciousness an ontological differentiating status. The limits of investigations into the human mind aren't the character of consciousness, but, plain and simply, the character of human science in general.

Neutral monism, understand from a pragmatic perspective, would, in accordance with my previous analysis and argumentation, be an acceptable philosophical alternative – even if still a partly metaphysical one – which gives an alternative interpretation of how perception and other mental processes occur. It is acceptable partly because it doesn't deny that the body is an essential part of the causal origin of basic perceptual contents that start meaningful mental processes, and, in this way, it helps current human science to explain not only the physical process of perception but also the link between this and first-personal experiences of it, making the latter increasingly less mysterious but no less important.[1]

---

[1]Some of the ideas presented in this paper were developed in two other papers: Stein, S.I.A (2013), "Naturalized phenomenalism and the talk of impressions", in *Mind, language and actions: papers of the 36th International Wittgenstein Symposium*, Kirchberg am Wechsel: Österreichische Ludwig Wittgenstein Gesellschaft, pp. 397-399; and Stein, S.I.A. (2016), "Mental elements as part of a scientific architecture", in Carmo, J.C. (ed.), *A companion to naturalism*, Pelotas (RS): Dissertatio Filosofia, pp. 217-228.

Bibliography

Chalmers, D.J. (1995). The puzzle of conscious experience. *Scientific American*, 273(6), December, 80-86. Updated version (2002) available at: http://consc.net/papers/puzzle.pdf, pp. 90-100.

——— (1996). *The conscious mind: in search of a fundamental theory*. New York: Oxford University Press.

Davidson, D. (2001). *Essays on actions and events*. Oxford: Clarendon.

Evans, G. (1982): *The varieties of reference*. Ed. J. McDowell. Oxford/New York: Claredon / Oxford University Press.

Feigl, H. ([1958] 1967): *The "mental" and the "physical": the essay and a postscript*. Minneapolis: University of Minnesota Press.

Levine, J. (1983). Materialism and qualia: the explanatory gap. *Pacific Philosophical Quarterly*, 64, 354-361.

Mach, E. ([1886] 1959). *The analysis of sensations*. 2nd ed. Transl. C.M. Williams & S. Waterlow. New York: Dover.

Maund, B. (2003). *Perception*. Chesham (GB): Acumen.

Quine, W.V. (1966). Russell's ontological development. *Journal of Philosophy*, 63(21), 657-667.

Russell, B. (1914a). On the nature of acquaintance. *The Monist*, 24, pp. 1-16, 161-87, 435-53. Reprinted in Russell, B. (1956). *Logic and knowledge*. Ed. R.C. Marsh. London: Routledge, pp. 125-174.

——— (1914b). The relation of sense-data to physics. *Scientia 16*, 1-27. Reprinted in Russell, B. (1918). *Mysticism and logic*. London: Longmans & Green, pp. 145-179.

——— (1919). On propositions: what they are and how they mean. *Proceedings of the Aristotelian Society*, supplementary volumes, *2*, 1-43. Reprinted in Russell, B. (1956). *Logic and knowledge*. Ed. R.C. Marsh. London: Routledge, pp. 283-320.

——— (1921). *The analysis of mind*. London: George Allen & Unwin.

——— (1927). *The analysis of matter*. London: Kegan Paul.

Sellars, W. ([1956] 1997). *Empiricism and the philosophy of mind*. Cambridge (MA): Harvard University Press.

Tully, R. (1988). Russell's neutral monism. *Russell: The Journal of the Bertrand Russell Archives*, 8(1), 209-224.

Chapter 5

The non-image-forming visual system: its relation to sleep, circadian rhythm, pupillary light reflex and other functions

Dora Fix Ventura
Dept. of Experimental Psychology, Institute of Psychology (DPE-IPUSP)
Núcleo de Apoio à Pesquisa em Neurociências e Comportamento (NAPNC-USP)
University of São Paulo (USP), São Paulo, SP, Brazil
dventura@usp.br

Ana Laura de Araújo Moura
DPE-IPUSP, NAPNC-USP, Dept. of Ophthalmology
Federal University of São Paulo (Unifesp), São Paulo, SP, Brazil

Gloria Liliana Duque-Chica
DPE-IPUSP, NAPNC-USP, University of Medellín, Medellín, Colombia

Balázs Vince Nagy
DPE-IPUSP, NAPNC-USP,
Dept. of Mechatronics, Optics, and Information Engineering
Budapest University of Technology and Economics, Budapest, Hungary

Carolina P.B. Gracitelli
Department of Ophthalmology
Federal University of São Paulo (Unifesp), São Paulo, SP, Brazil

Ana Luiza V. Milioni
DPE-IPUSP, NAPNC-USP

Kallene Vidal Summer
DPE-IPUSP, NAPNC-USP

Augusto Paranhos Jr.
Department of Ophthalmology
Federal University of São Paulo (Unifesp), São Paulo, SP, Brazil

Mirella Telles Salgueiro Barboni
DPE-IPUSP, NAPNC-USP

Abstract

The discovery of melanopsin in the retina provided answers to questions that had no response after decades of search. What controls the circadian rhythm of sleep and awakeness? Why do people have winter depression? The replies to these questions were surprising. A new type of cell, that is directly activated by light, as the cones and rods are, was found in the retina. This cell type forms a subset of retinal ganglion cells – the intrinsically photosensitive retinal ganglion cells (ipRGCs). They are also known as melanopsin-expressing retinal ganglion cells (mRGCs), after the pigment that is found in their soma and projections, which enables them to respond to light and is maximally activated by blue light, dominant in daylight. Research on the functions of these cells show that they give rise to a separate visual pathway, a non-image-forming visual pathway. Thus, in addition to the well-known image forming visual pathways, initiated by the activation of rods and cones, we have a second type of pathway initiated by the ipRGCs whose neural circuits are related to non-image-forming functions – pupillary light reflex, circadian rhythm, mood, and cognitive capacity. A brief review of these important discoveries and of our research on the topic is presented.

1. The discovery of a new visual system

A surprising discovery at the turn of the century showed that we have an additional visual system. Solid old knowledge establishes the existence, in humans and most vertebrates, of a duplex retina, giving origin to photopic and scotopic visual systems, activated respectively by cones and by rods. While cones are adapted to daylight vision, rods function at low light levels and mediate nocturnal vision. These two systems have been extensively studied and their function seemed to be sufficient to explain vision. This knowledge was shaken by the discovery that a third photoreceptor type existed in the mammalian retina (Provencio, Cooper & Foster, 1998; Provencio et al., 2000; Berson, 2003). Naturally, this had a great impact on visual science, more so because this new photoreceptor type is very different from the traditional types. Unlike rods and cones, the new photoreceptor is a ganglion cell, not a ciliated cell; its opsin is located over the entire cell, not just at the outer segment of the cell (Hattar et al., 2002, 2003; Dacey et al., 2005; Terakita, 2005; Isoldi et al., 2005). Due to its large size with ramifications spreading through large retinal areas, the new photoreceptor or the intrinsically photosensitive retinal ganglion cell (ipRGC), as it is called, is not seen as an image forming element of the visual system, as are rods and cones. It gives rise to an entirely different and previously unknown visual system, a non-image-forming visual system, which has other light dependent functions (Van Gelder, 2003).

The new system regulates circadian rhythms (Berson, 2003; Doyle & Menaker, 2007), the pupillary light reflex (Lucas et al., 2003; Hattar et al., 2002; Gamlin et al., 2007; Young & Kimura, 2008), and has several other functions (Turner & Mainster, 2008), like effect on mood dependent seasonal changes (LeGates et al., 2012; Cui et al., 2015), ambient light dependent responses in work environments or school (LeGates et al., 2012; Doyle & Menaker, 2007; Najjar et al., 2014), and migraine associated to light sensitivity (Noseda et al., 2010). This new cell also seems to be more resistant to injury than other ganglion cells (Cui et al., 2015; La Morgia et al., 2010; Hattar et al., 2002).

It is not surprising that the involvement of the non-image-forming system in so many functions will also have an impact on cognitive performance (Vandewalle, 2014). This is indeed the case for complex cognitive functions as shown in tasks involving working memory (Vandewalle et al., 2007, 2011; Chellappa et al., 2014), alertness and speed of information processing (Lehrl et al., 2007) or complex and simple reaction time (Schmoll et al., 2011), and in the age-related decrease of cognitive performance (Daneault et al., 2014; Schmoll et al., 2014).

This variety of functions requires specific neural substrates. Accordingly, the new system projects to several central nervous system structures that are in control of these functions (Hattar et al., 2002, 2006; Hannibal et al., 2014; Fernandez et al., 2016). Since these functions are very relevant to general health, it is now assumed that the eye has a critical role in well-being (Turner & Mainster, 2008). In addition to the non-image-forming functions, it is thought that the newly described system might also contribute to the image forming system, since projections to that system have been found (Hannibal et al., 2014; Dacey et al., 2005). Its contribution would be to signal ambient light level (Joo et al., 2013).

2. Neural substrate

The neural substrate of the non-image-forming visual system is based on the ipRGCs, the retinal ganglion cell subset type whose firing is light activated, also known as melanopsin-expressing retinal ganglion cells (mRGCs), after the pigment that is found in their soma and projections. The ipRGC is a rhabdomeric photopigment, homologous to and with the same signal transduction properties of invertebrate rhabdomeric photopigments (Panda et al., 2005; Isoldi et al., 2005; Berson, 2007; Mure et al., 2007, 2009; Xue et al., 2011; Provencio, Jiang et al., 1998). Melanopsin shares with invertebrate rhabdomeric photopigments the property of bi-stability: it is depleted by illumination with short wavelength light and regenerates when illuminated with long wavelength light (Mure et al., 2007, 2009; Melyan et al., 2005).

The characteristics of the ipRGC cell type as a photoreceptor have been studied in single cell recording experiments in which two main aspects called attention. First, it is sensitive to a wide band of wavelengths, in an analogous way to the absorption function of the opsins of rods and cones. Its opsin, called melanopsin, has a peak absorption in the blue spectral region, at 487 nm (Fu et al., 2005; Qiu et al., 2005). Second, it responds to light in a sustained way, suggesting that this type of response might have the function of signaling the information of the ambient light level (Dacey et al., 2005; Gamlin et al., 2007).

Light reaches the retina first at the ganglion cell layer – the rods and cones are the last layer of the retina reached by the incoming light. At the first layer, at the ganglion cell layer, the ipRGCs form a light sensitive surface. They are large and profusely branching cells. Melanopsin is found in the cell body and throughout all branches (Dacey et al., 2005). This is very different from the morphology of the rod and cone photoreceptors, in which the light sensitive part is a specialized part of the cell – the outer segment, and the remainder of the cell cannot be activated by light.

The ganglion cell subset formed by the ipRGCs leaves the retina as part of the optic nerve and projects to central structures that have been shown to be relevant to the pupillary light reflex and to circadian rhythms. The axons of the ipRGCs project to the olivary pretectal nucleus, which is part of the circuit of the pupillary light reflex. They also project to structures that regulate circadian rhythms – the suprachiasmatic nucleus, the subparaventricular zone, the ventrolateral preoptic area, and the intergeniculate leaflet (Benarroch, 2011; Morin, Blanchard & Provencio, 2003).

The change from daylight to darkness and back, informed by the ipRGCs to the higher centers, is seen as the mechanism that sets the circadian clock (Peirson, Halford & Foster, 2009). In addition to other factors, sleep disturbances may be affected by faulty functioning or absence of this day/night signaling mechanism, known as the zeitgeber of the circadian clock. It is therefore important to find out if ipRCG function is altered or not in patients with sleep complaints, especially in diseases that affect the optic nerve. In optic neuropathies, the loss of retinal ganglion cells may include loss of ipRGCs and thus constitute a sleep threatening condition. Patients with glaucoma, for example, are potential candidates for sleep disturbances and there is indeed a sleep complaint in this group (Gracitelli et al., 2015; Wang et al., 2013). On the other hand, in Leber's Hereditary Optic Neuropathy – a mitochondrial optic neuropathy believed to spare ipRGCs –

despite extensive ganglion cell loss, patients do not refer sleep problems (Moura et al., 2013; La Morgia et al., 2010).

3. Pupillary light reflex

The integrity of the ipRGCs may be assessed non-invasively by the pupillary light reflex (PLR). When light is presented, pupil diameter immediately decreases and then returns to an intermediate size until light is turned off, allowing it to return to the baseline diameter. This "escape response", a sustained constriction, has been attributed to the direct response to light of the ipRGCs, while the initial transient peak response is mediated by rod and cone inputs to the same ipRGCs (Kardon et al., 2009; Gamlin et al., 2007).

The selection of appropriate stimuli allows separation of these two input types into the resulting PLR – rod and cone photoreceptors and ipRGC direct response to light. Wavelength of the stimulation light in the blue spectral region, around 487 nm, activates ipRGCs maximally since this is the peak of the melanopsin absorption curve. L- and M-cones, on the other hand may be activated by light stimuli in the long wavelength region without disturbing the ipRGCs. The "winner takes all" mechanism described by McDougal & Gamlin (2010) postulates the relative contributions of inner and outer retina to PLR. Contributions of rods and cones determine the spectral sensitivity of the PLR, if a stimulus is under ipRGCs threshold. In a parametric study designed to define the most effective conditions for a clinically useful protocol, Park et al. (2011) compared PLRs to a 1 s blue (470 nm) light versus those in response to red (640 nm) lights, having arrived at the following stimulus parameters: low intensity blue flash to study rod mediated responses, red or blue flashes with blue background for cone mediated responses, and high intensity red and blue flashes for ipRGC mediated responses.

Fig. 5.1 illustrates pupillary light responses to 1 s high intensity blue and red flashes, showing that the transient peak response (maximal constriction) has the same amplitude for both, but the sustained response is very different. The blue flash causes a long sustained constriction while with the red flash the pupillary diameter returns to baseline much faster. The melanopsin of ipRGC response is represented by the difference in the amplitude of the sustained response to the red and blue flashes. The parameters defined by Park and coworkers proved useful in further studies (Moura et al., 2013; Gracitelli et al., 2015, 2016; Duque-Chica et al., 2017) and have been systematically used by our group to investigate a number of pathological conditions.

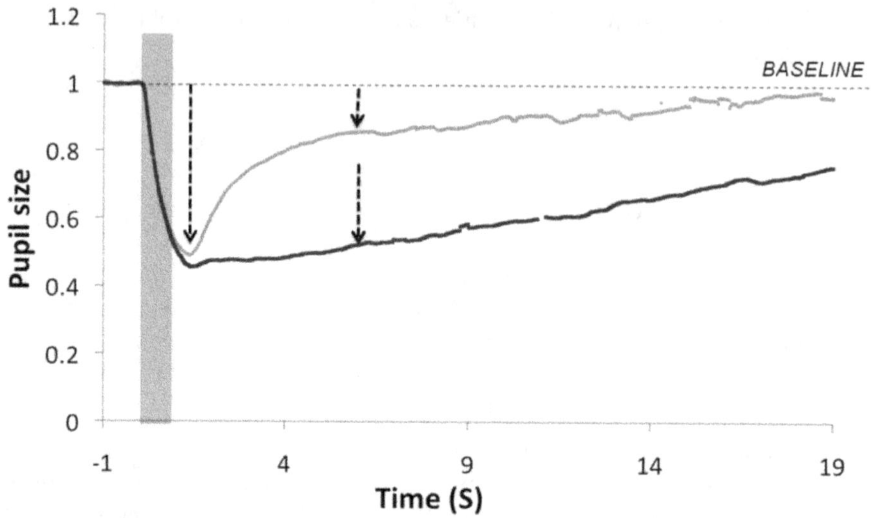

Figure 5.1: Temporal course of the pupillary light response (PLR) to 1 s blue (black trace) and red (grey trace) flashes of high intensity, shown as relative pupillary diameter as a function of time. After the flash, there is a fast constriction leading to a peak, followed by a reduction in the constriction (sustained response). And then, the pupil diameter returns to the baseline level. (Figure prepared by the authors, previously unpublished.)

4. Clinical applications

4.1 PLR – Leber's hereditary optic neuropathy (LHON)

Our research group studied the ipRGC responses using the PLR from a group of Leber's hereditary optic neuropathy (LHON) patients belonging to a giant pedigree found in Brazil (Sadun et al., 2003). LHON is a hereditary disease that causes severe loss of vision. It is due to a mutation in the mitochondrial DNA. Patients become affected around 15 to 40 years of age, by a sudden and devastating wave of ganglion cell apoptosis, first in one eye, a few months later in the other eye. Although there is massive ganglion cell death reducing the ganglion cell population to 10%, the ipRGCs are not so strongly affected since its population suffers only a 50% loss, suggesting a relative preservation of these cells (La Morgia et al., 2010). Accordingly, clinical reports do not describe PLR alterations and patients do not refer sleep disturbances.

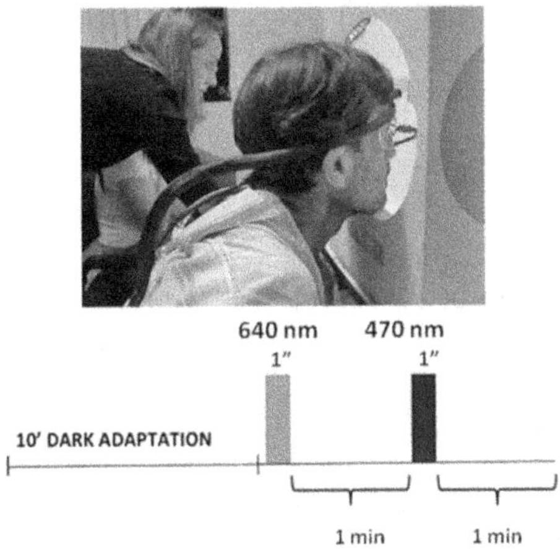

Figure 5.2: Top part shows subject placed at the Ganzfeld stimulus with light turned on. Bottom part shows stimulation protocol: after 10 min of dark adaptation 1 s red (640 nm) and blue (470 nm), flashes are presented with a 1 min interval between them. The flash presentation sequence is repeated three times. (Figure prepared by the authors, previously unpublished.)

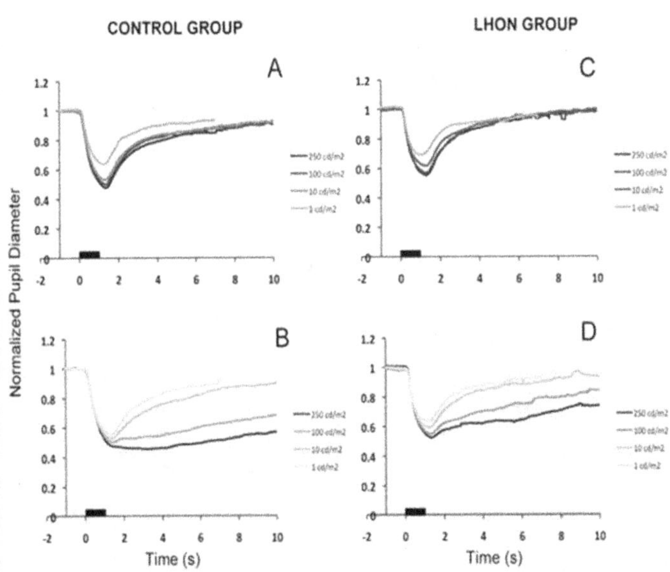

Figure 5.3: PLR traces obtained with red (A, C) and blue (B, D) flashes are shown as normalized pupil diameter as a function of time, in controls (A, B) and LHON patients (C, D). (Modified from Moura et al., 2013, p. 4473; permission for publication granted by *Investigative Ophthalmology & Visual Science*).

The red and blue flash stimuli were presented to the patients and controls in a Ganzfeld while the PLR was recorded as pupil diameter (Fig. 5.2). PLR traces in response to stimuli of different intensities are superimposed showing results from patients and controls to red and blue flashes (Fig. 5.3). Overall the differences between patients and controls seem small. Fig. 5.4 shows a quantitative description of the data by displaying pupil constriction amplitude as a function of flash luminance for both the transient response, due to input from rods and cones and the sustained response, measured at 6 s after flash onset. Red and blue symbols correspond respectively to responses to the red and blue flashes.

The transient PLR is 15% smaller in LHON patients compared to controls ($p < .05$), but five out of ten patients had amplitude within the control range. The sustained PLR was comparable to that of controls at low intensities but a 27% loss was observed at the highest intensity, where the ipRGC response is most obvious. Here also there was considerable overlap with control values. We concluded that melanopsin responses were indeed present in LHON patients and this explained why they did not have sleep complaints (Moura et al., 2013).

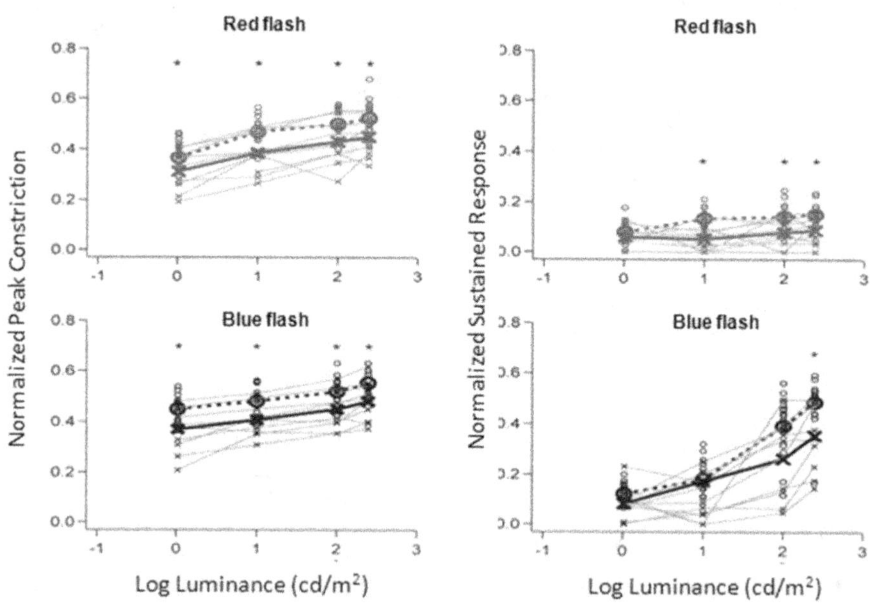

Figure 5.4: The left column shows averages of the peak response amplitude to red and blue flashes for controls (circles) and patients (crosses). Top graphs show measurements of the PLR to red flashes, bottom graphs to blue flashes. The right column shows the same for the sustained response (pupil diameter measured 6 s after flash presentation). Asterisks indicate significant differences between LHON and controls. (Modified from Moura et al., 2013, p. 4474; permission for publication granted by *Investigative Ophthalmology & Visual Science*).

4.2 PLR – Glaucoma

A much more obvious connection between ipRGC activity seen in the PLR and sleep disturbances characterized a group of glaucoma patients, which constituted a clinical population with a different optic neuropathy compared to LHON. This group had a clear association between the average retinal nerve fiber layer (RNFL) thickness, measured by optic tomography (Cirrus HD-OCT) and the sustained PLR to the high intensity blue flash ($p < .024$) (Gracitelli et al., 2014). Their decreased ipRGC response was associated with a general decrease of sleep quality measured by polysomnography parameters (Gracitelli et al., 2015) and also with daytime sleepiness, as assessed by a sleep scale (Epworth Sleepiness Scale) (Gracitelli et al., 2016) (Fig. 5.5). The main clinical finding of these studies performed by our research group were that glaucoma patients had increased daytime sleepiness compared with control subjects, and these results were associated with worse polysomnographic parameters and lower sustained PLRs to a blue flash. These major non-image forming functions of ipRGCs should be routinely evaluated in certain patients with glaucoma, since excessive daytime sleepiness and worse parameters in polysomnography are well known to affect quality of life, daytime function, and mortality.

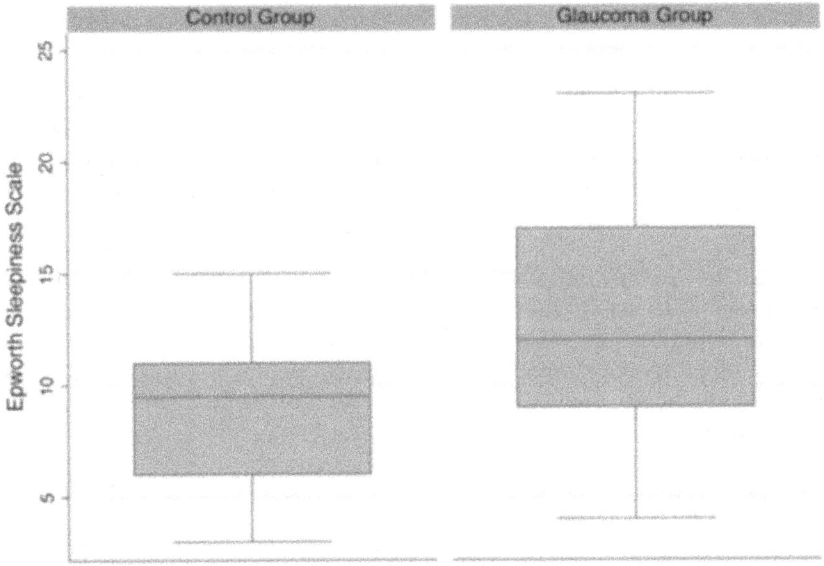

Figure 5.5: Epworth Sleepiness Scale scores in control and glaucoma groups. (Gracitelli et al., 2016, p. 4; permission for publication granted by the *Journal of Ophthalmology*).

Furthermore, in this glaucoma patient group the patients were ranked as to glaucoma severity and arranged in subgroups of mild, moderate and severe glaucoma (according to the Hoddap Parish Anderson classification). We found that the PLR mediated by the outer retina photoreceptors – rods and cones – was altered in glaucoma, in addition to the ipRGC mediated PLR (inner retina photoreceptors). These effects occurred in moderate and severe glaucoma, but not in mild glaucoma. In addition, a significant loss of color discrimination in the tritan (blue) axis was observed in all groups (Duque-Chica et al., 2017).

4.3 Cognitive functions

In view of the characteristic of bistability of melanopsin, confirmed in electrophysiological recordings in animal studies (Mure et al., 2007) and in human PLR results (Mure et al., 2009), we examined the effects of ambient light spectral content in different cognitive abilities of healthy volunteers. Melanopsin activation and deactivation used blue and amber LED illumination with similar irradiance levels, while performance of participants was assessed in neuropsychological cognitive tests. Results suggested that the specific color of pre-stimulation might result in differences in cognitive performance (Nagy et al., 2015). These results confirmed earlier findings on alertness and speed of information processing obtained with similar experimental design (Lehrl et al., 2007).

5. Summary, conclusions and impact

The discovery of a non-image-forming visual system had an unprecedented impact in vision science showing that a very broad range of behaviors is regulated and may be affected by the activity of the system that has as its main function the signaling of day and night. This system is important to very vital functions, so much so that the health considerations involved place the eye as an organ relevant to general health. Recent developments of non-invasive techniques made it possible to access the activity of the non-image-forming visual system through a common eye response, the pupillary light reflex, allowing the estimation of its integrity. We have found these techniques to be reliable and useful to characterize populations whose optic nerves have been affected by different diseases.[1]

[1] This work was supported by grants to DFV from FAPESP (Fundação de Amparo à Pesquisa do Estado de São Paulo), Projeto Temático 08/58731-2 and 14/26818-2. DFV is a CNPq (National Council for Scientific and Technological Development) research fellow; BVN had a FAPESP Visiting Professorship 2014/06457; KVS is a FAPESP 2015/22227-2 postdoctoral fellow at DFV's lab; CPBG had a doctoral fellowship from CAPES (Brazilian Coordination for the Improvement of Higher Education Personnel), 12309-13-3.

Bibliography

Benarroch, E.E. (2011). The melanopsin system: phototransduction, projections, functions, and clinical implications. *Neurology*, 76(16), 1422-1427.

Berson, D.M. (2003). Strange vision: ganglion cells as circadian photoreceptors. *Trends in Neurosciences*, 26(6), 314-320.

——— (2007). Phototransduction in ganglion-cell photoreceptors. *Pflügers Archiv -European Journal of Physiology*, 454(5), 849-855.

Chellappa, S.L.; Ly, J.Q.M.; Meyer, C.; Balteau, E.; Degueldre, C.; Luxen, A.; Phillips, C.; Cooper, H.M. & Vandewalle, G. (2014). Photic memory for executive brain responses. *Proceedings of the National Academy of Sciences U.S.A.*, 111(16), 6087-6091.

Cui, Q.; Ren, C.; Sollars, P.J.; Pickard, G.E. & So, K.F. (2015). The injury resistant ability of melanopsin-expressing intrinsically photosensitive retinal ganglion cells. *Neuroscience*, 284, 845-853.

Dacey, D.M.; Liao, H.W.; Peterson, B.B.; Robinson, F.R.; Smith, V.C.; Pokorny, J.; Yau, K.W. & Gamlin, P.D. (2005). Melanopsin-expressing ganglion cells in primate retina signal colour and irradiance and project to the LGN. *Nature*, 433 (7027), 749-754.

Daneault, V.; Hebert, M.; Albouy, G.; Doyon, J.; Dumont, M.; Carrier, J. & Vandewalle, G. (2014). Aging reduces the stimulating effect of blue light on cognitive brain functions. *Sleep*, 37(1), 85-96.

Doyle, S. & Menaker M. (2007). Circadian photoreception in vertebrates. *Cold Spring Harbor Symposia on Quantitative Biology*, 72, 499-508.

Duque-Chica, G.L.; Gracitelli, C.P.B.; Moura, A.L.A.; Nagy, B.V.; Vidal, K.S.; Delegrego, P.; Paranhos Jr., A. & Ventura, D.F. (2017). The contribution of inner and outer retinal photoreceptors to the pupil light response in glaucomatous optic neuropathy. Submitted.

Fernandez, D.C.; Chang, Y.T.; Hattar, S. & Chen, S. K. (2016). Architecture of retinal projections to the central circadian pacemaker. *Proceedings of the National Academy of Sciences U.S.A.*, 113(21), 6047-6052.

Fu, Y.B.; Zhong, H.N.; Wang, M.H.H.; Luo, D.G.; Liao, H.W.; Maeda, H.; Hattar, S.; Frishman, L.J. & Yau, K.W. (2005). Intrinsically photosensitive retinal ganglion cells detect light with a vitamin A-based photopigment, melanopsin. *Proceedings of the National Academy of Sciences U.S.A.*, 102 (29), 10339-10344.

Gamlin, P.D.R.; McDougal, D.H.; Pokorny, J.; Smith, V.C., Yau, K.W. & Dacey, D. M. (2007). Human and macaque pupil responses driven by melanopsin-containing retinal ganglion cells. *Vision Research*, 47(7), 946-954.

Gracitelli, C.P.B.; Duque-Chica, G.L.; Moura, A.L.A.; Nagy, B.V.; Ragot de Melo, G.; Roizenblatt, M.; Borba, P.D.; Teixeira, S.H.; Ventura, D.F. & Paranhos Jr., A. (2014). A positive association between intrinsically photosensitive retinal ganglion cells and retinal nerve fiber layer thinning in glaucoma. *Investigative Ophthalmology and Visual Science*, 55(12), 7997-8005.

Gracitelli, C.P.B.; Duque-Chica, G.L.; Moura, A.L.A.; Roizenblatt, M.; Nagy, B.V.; Ragot de Melo, G.; Borba, P.D.; Teixeira, S.H.; Tufik, S.; Ventura, D.F. & Paranhos Jr., A. (2016). Relationship between daytime sleepiness and intrinsically photosensitive retinal ganglion cells in glaucomatous disease. *Journal of Ophthalmology*, article 5317371, 9 pp.

Gracitelli, C.P.B.; Duque-Chica, G.L.; Roizenblatt, M.; Moura, A.L.A.; Nagy, B.V.; Ragot de Melo, G.; Borba, P.D.; Teixeira, S.H.; Tufik, S.; Ventura, D.F. & Paranhos Jr., A. (2015). Intrinsically photosensitive retinal ganglion cell activity is associated with decreased sleep quality in patients with glaucoma. *Ophthalmology*, 122(6), 1139-1148.

Hannibal, J.; Kankipati, L.; Strang, C.E.; Peterson, B.B.; Dacey, D. & Gamlin, P.D. (2014). Central projections of intrinsically photosensitive retinal ganglion cells in the macaque monkey. *Journal of Comparative Neurology*, 522(10), 2231-2248.

Hattar, S.; Kumar, M.; Park, A.; Tong, P.; Tung, J.; Yau, K.W. & Berson, D.M. (2006). Central projections of melanopsin-expressing retinal ganglion cells in the mouse. *Journal of Comparative Neurology*, 497(3), 326-349.

Hattar, S.; Liao, H.W.; Takao, M.; Berson, D.M. & Yau, K.W. (2002). Melanopsin-containing retinal. ganglion cells: architecture, projections, and intrinsic photosensitivity. *Science*, 295 (5557), 1065-1070.

Hattar, S.; Lucas, R.J.; Mrosovsky, N.; Thompson, S.; Douglas, R.H.; Hankins, M.W.; Lem, J.; Biel, M.; Hofmann, F.; Foster, R.G. & Yau, K.W. (2003). Melanopsin and rod-cone photoreceptive systems account for all major accessory visual functions in mice. *Nature*, 424 (6944), 76-81.

Isoldi, M.C.; Rollag, M.D.; Castrucci, A.M.D. & Provencio, I. (2005). Rhabdomeric phototransduction initiated by the vertebrate photopigment melanopsin. *Proceedings of the National Academy of Sciences U.S.A.*, 102(4), 1217-1221.

Joo, H.R.; Peterson, B.B.; Dacey, D.M.; Hattar, S. & Chen, S.K. (2013). Recurrent axon collaterals of intrinsically photosensitive retinal ganglion cells. *Visual Neuroscience*, 30(4), 175-182.

Kardon, R.; Anderson, S.C.; Damarjian, T.G.; Grace, E.M.; Stone, E. & Kawasaki, A. (2009). Chromatic pupil responses preferential activation of the melanopsin-mediated versus outer photoreceptor-mediated pupil light reflex. *Ophthalmology*, 116(8), 1564-1573.

La Morgia, C.; Ross-Cisneros, F.N.; Sadun, A.A.; Hannibal, J.; Munarini, A.; Mantovani, V.; Barboni, P.; Cantalupo, G.; Tozer, K.R.; Sancisi, E.; Salomão, S.R.; Moraes, M.N.; Heegaard, S.; Milea, D.; Kjer, P.; Montagna, P. & Carelli, V. (2010). Melanopsin retinal ganglion cells are resistant to neurodegeneration in mitochondrial optic neuropathies. Brain, 133, 2426-2438.

LeGates, T.A.; Altimus, C.M.; Wang, H.; Lee, H.K.; Yang, S.G.; Zhao, H.Q.; Kirkwood, A.; Weber, E.T. & Hattar, S. (2012). Aberrant light directly impairs mood and learning through melanopsin-expressing neurons. *Nature*, 491 (7425): 594-598.

Lehrl, S.; Gerstmeyer, K.; Jacob, J.H.; Frieling, H.; Henkel, A.W.; Meyrer, R.; Wiltfang, J.; Kornhuber, J. & Bleich, S. (2007). Blue light improves cognitive performance. *Journal of Neural Transmission*, 114(4), 457-460.

Lucas, R.J.; Hattar, S.; Takao, M.; Berson, D.M.; Foster, R.G. & Yau, K.W. (2003). Diminished pupillary light reflex at high irradiances in melanopsin-knockout mice. *Science*, 299 (5604), 245-247.

McDougal, D.H. & Gamlin, P. D. (2010). The influence of intrinsically-photosensitive retinal ganglion cells on the spectral sensitivity and response dynamics of the human pupillary light reflex. *Vision Research*, 50(1), 72-87.

Melyan, Z.; Tarttelin, E.E.; Bellingham, J.; Lucas, R.J. & Hankins, M.W. (2005). Addition of human melanopsin renders mammalian cells photoresponsive. *Nature*, 433 (7027), 741-745.

Morin, L.P.; Blanchard, J.H. & Provencio, I. (2003). Retinal ganglion cell projections to the hamster suprachiasmatic nucleus, intergeniculate leaflet, and visual midbrain: bifurcation and melanopsin immunoreactivity. *Journal of Comparative Neurology*, 465(3), 401-416.

Moura, A.L.A.; Nagy, B.V.; La Morgia, C.; Barboni, P.; Oliveira, A.G.F.; Salomão, S.R.; Berezovsky, A.; de Moraes-Filho, M.N.; Chicani, C.F.; Belfort Jr., R.; Carelli, V.; Sadun, A.A.; Hood, D.C. & Ventura, D.F. (2013). The pupil light reflex in Leber's hereditary optic neuropathy: evidence for preservation of melanopsin-expressing retinal ganglion cells. *Investigative Ophthalmology & Visual Science*, 54(7), 4471-4477.

Mure, L.S.; Cornut, P.L.; Rieux, C.; Drouyer, E.; Denis, P.; Gronfier, C. & Cooper, H.M. (2009). Melanopsin bistability: a fly's eye technology in the human retina. *PLOS ONE*, 4(6), e5991.

Mure, L.S.; Rieux, C.; Hattar, S. & Cooper, H.M. (2007). Melanopsin dependent nonvisual responses: Evidence for photopigment bistability in vivo. *Journal of Biological Rhythms*, 22(5), 411-424.

Najjar, R.P.; Wolf, L.; Taillard, J.; Schlangen, L.J.M.; Salam, A.; Cajochen, C. & Gronfier, C. (2014). Chronic artificial blue-enriched white light is an effective countermeasure to delayed circadian phase and neurobehavioral decrements. *PLOS ONE*, 9(7), e102827.

Noseda, R.; Kainz, V.; Jakubowski, M.; Gooley, J.J.; Saper, C.B.; Digre, K. & Burstein, R. (2010). A neural mechanism for exacerbation of headache by light. *Nature Neuroscience*, 13(2), 239-245.

Panda, S.; Nayak, S.K.; Campo, B.; Walker, J.R.; Hogenesch, J.B. & Jegla, T. (2005). Illumination of the melanopsin signaling pathway. *Science*, 307 (5709), 600-604.

Park, J.C.; Moura, A.L.A.; Raza, A.S.; Rhee, D.W.; Kardon, R.H. & Hood, D.C. (2011). Toward a clinical protocol for assessing rod, cone, and melanopsin contributions to the human pupil response. *Investigative Ophthalmology & Visual Science*, 52(9), 6624-6635.

Peirson, S.N.; Halford, S. & Foster, R.G. (2009). The evolution of irradiance detection: melanopsin and the non-visual opsins. *Philosophical Transactions of the Royal Society B*, 364 (1531), 2849-2865.

Provencio, I.; Cooper, H.M. & Foster, R.G. (1998). Retinal projections in mice with inherited retinal degeneration: implications for circadian photoentrainment. *Journal of Comparative Neurology*, 395(4), 417-439.

Provencio, I.; Jiang, G.S.; De Grip, W.J.; Hayes, W.P. & Rollag, M.D. (1998). Melanopsin: an opsin in melanophores, brain, and eye. *Proceedings of the National Academy of Sciences U.S.A.*, 95(1), 340-345.

Provencio, I.; Rodriguez, I.R.; Jiang, G.S.; Hayes, W.P.; Moreira, E.F. & Rollag, M.D. (2000). A novel human opsin in the inner retina. *Journal of Neuroscience*, 20 (2), 600-605.

Qiu, X.D.; Kumbalasiri, T.; Carlson, S. M.; Wong, K.Y.; Krishna, V.; Provencio, I. & Berson, D.M. (2005). Induction of photosensitivity by heterologous expression of melanopsin. *Nature*, 433 (7027), 745-749.

Sadun, A.A.; Carelli, V.; Salomão, S.R.; Berezovsky, A.; Quiros, P.A.; Sadun, F.; DeNegri, A.M.; Andrade, R.; Moraes, M.; Passos, A.; Kjaer, P.; Pereira, J.; Valentino, M.L.; Schein, S. & Belfort, R. (2003). Extensive investigation of a large Brazilian pedigree of 11778/haplogroup J Leber hereditary optic neuropathy. *American Journal of Ophthalmology*, 136(2), 231-238.

Schmoll, C.; Khan, A.; Aspinall, P.; Goudie, C.; Koay, P.; Tendo, C.; Cameron, J.; Roe, J.; Deary, I. & Dhillon, B. (2014). New light for old eyes: comparing melanopsin-mediated non-visual benefits of blue-light and UV-blocking intraocular lenses. *British Journal of Ophthalmology*, 98(1), 124-128.

Schmoll, C.; Tendo, C.; Aspinall, P. & Dhillon, B. (2011). Reaction time as a measure of enhanced blue-light mediated cognitive function following cataract surgery. *British Journal of Ophthalmology*, 95(12), 1656-1659.

Terakita, A. (2005). The opsins. *Genome Biology*, 6(3), article 213, 9 pp.

Turner, P.L. & Mainster, M.A. (2008). Circadian photoreception: ageing and the eye's important role in systemic health. *British Journal of Ophthalmology*, 92(11), 1439-1444.

Van Gelder, R.N. (2003). Making (a) sense of non-visual ocular photoreception. *Trends in Neurosciences*, 26(9), 458-461.

Vandewalle, G. (2014). La lumière comme stimulant de l'activité cognitive cérébrale. *Médecine Sciences*, 30(10), 902-909.

Vandewalle, G. ; Archer, S.N.; Wuillaume, C.; Balteau, E.; Degueldre, C.; Luxen, A.; Dijk, D.J. & Maquet, P. (2011). Effects of light on cognitive brain responses depend on circadian phase and sleep homeostasis. *Journal of Biological Rhythms*, 26(3), 249-259.

Vandewalle, G.; Gais, S.; Schabus, M.; Balteau, E.; Carrier, J.; Darsaud, A.; Sterpenich, V.; Albouy, G.; Dijk, D.J. & Maquet, P. (2007). Wavelength-dependent modulation of brain responses to a working memory task by daytime light exposure. *Cerebral Cortex*, 17(12), 2788-2795.

Wang, H.Z.; Zhang, Y.; Ding, J.M. & Wang, N.L. (2013). Changes in the circadian rhythm in patients with primary glaucoma. *PLOS ONE*, 8(4), e62841.

Xue, T.; Do, M.T.H.; Riccio, A.; Jiang, Z.; Hsieh, J.; Wang, H.C.; Merbs, S.L.; Welsbie, D.S.; Yoshioka, T.; Weissgerber, P.; Stolz, S.; Flockerzi, V.; Freichel, M.; Simon, M.I.; Clapham, D.E. & Yau. K.W. (2011). Melanopsin signalling in mammalian iris and retina. *Nature*, 479 (7371), 67-73.

Young, R.S.L. & Kimura, E. (2008). Pupillary correlates of light-evoked melanopsin activity in humans. *Vision Research*, 48(7), 862-871.

Chapter 6

Can brain connectivity analysis help understand brain functions?

Koichi Sameshima
Medical School
University of São Paulo (USP), São Paulo, SP, Brazil
ksameshi@usp.br

Luiz A. Baccalá
Polytechnic School
University of São Paulo (USP), São Paulo, SP, Brazil
baccala@lcs.poli.usp.br

Abstract

The brain has long been accepted as functionally segregated and hierarchically organized, so that one of neuroscience's important experimental paradigms has been the study of how neurons (or more generally neural structures) interact to process information and generate behavior. In the 1970's and 1980's, while studying cooperativity and interaction between pairs of neurons, Gerstein and colleagues coined the term *effective connectivity*, which encompasses today's popular notions of functional and effective connectivity. The introduction of the Granger causality concept and its many related representations and extensions to the multivariate and frequency domains has allowed inferring the interaction of tens of simultaneously analyzed neural structures. Many practical challenges for the application of the connectivity inference methods remain. In fact, even assuming technical limitations are eventually overcome to precisely expose the true dynamic connectivity between neural structures, one must ask what this complete description of connectivity can actually reveal about brain functional mechanisms. No definitive verdict is available yet. Ultimately, we think that larger scale clinical/experimental trials to validate analysis techniques for each type of connectivity description are required. Reliably determining the where and the how of brain function remains challenging and calls for multidisciplinary approaches to integrate and validate data from different sources.

A mathematician, like a painter or a poet, is a maker of patterns. If his patterns are more permanent than theirs, it is because they are made with ideas.
(G.H. Hardy, 1940, p. 84)

1. Introduction

Broadly some four somewhat overlapping functions, (a) homeostasis, (b) genesis of behavior, (c) learning and memory, and (d) information processing, may be attributed to the brain, in our quest to understand how it works, performs cognitive functions and generates behavior.

In the preface of his *The handbook of brain theory and neural networks* (1995), Michael A. Arbib confesses that two questions inspired the volume: (1) "How does the brain work?", and (2) "How can we build intelligent machines?". He goes on to say: "It provides no simple, single answer to either question because no single answer, simple or otherwise, exists. However, in hundreds of articles it charts the immense progress made in recent years in answering many related, but far more specific, questions" (Arbib, 1995, p. xi). In its 266 articles, the book recounts brain theories and neural networks as understood then, without even a single reference to either *connectivity analysis* or *causality*. Whereas the second question above has enjoyed enormous progress under the *artificial intelligence* heading – with the development of tools that faithfully mimic cognitive capacities hitherto considered exclusively human, like speech recognition, language comprehension and translation, car driving, and playing complex board games such as the chess and the game of go – little has come of it in helping to explain the brain's inner working. In fact, one might ask oneself, what enlightenment can bring feats, like the much publicized 2016 AlphaGo system's (developed by the Google-DeepMind Technologies) victory over the world's best human professional go player, towards the first question above?

Brain studies, perhaps due to their complexity and wide recognition as being at today's frontier of science, have attracted researchers beyond medicine or biology, to include mathematicians, physicists, engineers, and philosophers. Despite all this massive engagement, it is amazing and even somewhat ironic to realize that the bulk of neuroscience's accumulated knowledge has come from experiments in model organisms as diverse as worms, ants, bees, flies, jellyfishes, aplysias, squids, birds, mouse, rats, cat, dog, dolphins, monkeys and human, to list just a few of the animal species from which morphology, biochemistry, anatomy, and nervous system physiology and animal behavior have been studied.

In the case of mammals, neuroscientists deal with neural structures that encompass several orders of magnitude both in space (from the molecules to brains) and in time, some 12 orders of magnitude from the milliseconds of action potential duration to the hundred-odd years in the lifespan of some species. The issue of time and spatial scales is important in neurophysiological phenomena, since its resolution is very much constrained by the measurement technologies and experimental protocols used in probing neural structures and neuronal activity (see Fig. 6.1). From the functional assessment perspective, focusing on common building blocks and organizational aspects that constitute the nervous system has proven fruitful in understanding the human brain, allowing the systematic and comparative collection of its data with respect to information from insects, mollusks, reptiles, amphibians, and various vertebrates.

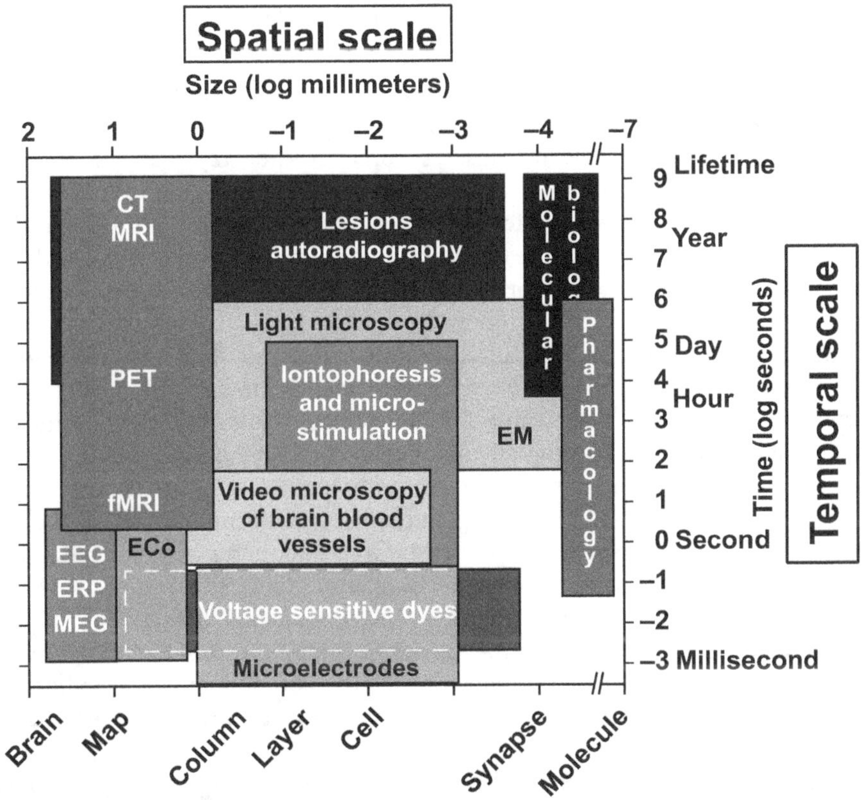

Figure 6.1: The diagram shows the spatial and temporal scales in which the brain can be explored, depicting the arrays of available technologies to access brain functional mappings at the intersection of a specific spatial scale with a given temporal resolution. For instance, to look at the cell level action potential on a millisecond scale, one needs microelectrodes recording either single/multi-unit action potentials or local field potentials, or voltage sensitive dyes method. CT: computed tomography; MRI: magnetic resonance imaging; PET: positron emission tomography; fMRI: functional MRI; EEG: electroencephalography; ERP: event-related potential; MEG: magnetoencephalography; Eco: electrocorticography; EM: electron microscopy. (Modified from Posner & Raichle, 1994, p. 14).

The neuron is the fundamental information processing element of the nervous system. It is made up by a main cell body with its many dendrites and an axon that projects itself, reaching other neurons via synapses to their dendrites or cell bodies, using a biochemical machinery of receptor molecules, neurotransmitters, ion channels, and plasticity mechanisms that, interestingly, seems to be common and similar across species.

For over more than a hundred years neuroscientists have mostly employed *divide and conquer* strategies for studying these spatial and temporal scales with tools from genetic, molecular, cellular, anatomical, cognitive and behavioral aspects of neural systems and animal models. The mainstream approach has been to fill the large and complex mosaic of the central nervous system and of the brain piece by piece.

Early claims (phrenologist allegations) that externally examining of mammalian brain shape in the form of lobes and sulci could explain how it works have long been totally dispelled. Despite this, as some general key features of brain organization the vertebrate brains do display modularity into functionally segregated brain structures, in addition to the integration of specialized areas that together allow delimiting specific functions that can be assigned to different brain areas. Motor and primary sensory areas of vision, touch and hearing are good examples, even though we now know, from recent plasticity studies in awake animals and clues from phantom limb patients, that the functional representations continuously and dynamically change in size and location from second to second and over whole lifespan scales (Ramachandran & Hirstein, 1998). Brain modularity was initially discovered by studying cognitive and motor deficits observed in patients or soldiers with brain lesions or tumors. More recently these attempts at discovering and rediscovering functionally segregated brain areas have flourished with the widespread adoption of brain functional mapping methods using diverse technologies (old and new), such as electrophysiological recording, electroencephalography (EEG) magnetoencephalography (MEG), positron emission tomography (PET), functional magnetic resonance imaging (fMRI), and functional near-infrared spectroscopy (fNIRS). These techniques allow looking at the activated brain areas under peripheral sensory stimulation and even during the performance of behavioral tasks, and many of them – EEG, MEG, fMRI – have the added advantage of being noninvasive, which allows repeated measurements.

One good example of this was the discovery of the BOLD (Blood Oxygen Level Dependent) contrast by Ogawa et al. (1990), which led to the immediate development of fMRI (Functional Magnetic Resonance Imaging) (Kwong et al., 1992), where multi-slice brain images are collected every few seconds via the so-'called T2* contrast, that measures the difference in magnetic properties between oxyhemoglobin and deoxyhemoglobin, pointing to local increases in oxygen level close to active cerebral areas that can thus be visualized as small signal increases of the order of 1 to 5%. The technique has become ubiquitous and enjoys enormous success in highlighting functionally specialized brain areas for systems and cognitive neuroscience studies, despite frequent overinterpretations commonly referred to as *neo-phrenology* (Friston, 2002). Despite worries about reliable fMRI detection, performed at a voxel by voxel basis over the whole brain, it is universally recognized that brain activations do not occur in isolation, i.e. brain responses often involve interactions between a complex network of interconnected regions, that may reflect more fundamental brain functional architectures and principles that could differentially characterize normal and pathological brain dynamics, therefore being of great potential clinical relevance.

2. From where to how

The last three decades has seen a methodological shift in the characterization of brain dynamics, from the early prevalent searches for identifying areas with increased neural activity under stimulus or involved in some particular brain function focused on the essentially phrenological *where* question (the neo-phrenology issue), toward the quest of answering *how* brain areas interact and integrate information (functional integration) during the enactment of cognitive tasks or over the genesis of behavior. The general term 'connectivity' or 'causality analysis' has been liberally used to describe this trend. Early on, however, the need was felt for qualifying what one precisely means by it. Both clinical and basic neuroscientists, using multichannel local field potential (LFP) recording, fMRI, EEG and MEG, remain in need of clear and straightforward tools to determine causal relationships among brain structures.

Nowadays, it is easy to do brain map imaging experiments. Not so long ago, recording was restricted to no more than one-channel single-unit stable signal with microelectrode measurements in anesthetized animals, in what was a truly daunting task. To assess connectivity between a pair of neurons in the mammalian cortex while doing controlled evoked stimulus is a difficult task since one must be certain that the areas of interest are anatomically interconnected. This happens, for example, in the case of a pair of neurons only a millimeter apart, where the electrophysiologist inserts a pair of microelectrode tips using a micromanipulator and stereotactic system, while fighting to keep them spatially stable, sometimes for tens of minutes to hours. There were several improvements in both electrode technology and in analysis methods, so that today it is possible to sample tens or hundreds of single- or multi-unit neuronal activities. One such technology, that improved sorting-out the single-unit neuronal activity, was the invention of stereotrode by McNaughton, O'Keefe & Barnes (1983), with the idea of sensing one neuron with two or more microelectrodes. Today the most popular stereotrode is the tetrode, with spike-sorting software built into many commercial equipment, in which four microwire microelectrodes are put together with exposed tips as close as 20 to 50 μm (McNaughton et al., 1983; Gray et al., 1995).

In the 1980s and 1990s, the prevailing approaches for studying neuron cooperativity and interaction were via (1) the cross-correlation of simultaneously recorded action potential activities during an appropriate experimental condition, and (2) the joint peri-stimulus time histogram (JPSTH), which corresponds to a stimulus-locked time decomposition of cross-correlation (see Gerstein & Perkel, 1969, 1972). Analyzing the presence or absence of the peak of cross-correlograms and its possible relative delays enables one to infer, on a pair-wise neuron analysis basis, whether (a) they share a common input, or if (b) one of them influences the other, with the relative delay magnitude providing information as to whether the connection is direct or through other neurons (considering typical synaptic delay). Under the latter scenarios, in which the interaction could be experimentally evidenced, Gerstein and colleagues coined the term *effective connectivity* (Aertsen & Gerstein, 1991). With it they meant to indicate the presence of either statistically significant coactivation, sharing common input, or direct or indirect interaction between pairs of neurons under scrutiny as measured through cross-correlograms. This definition carries a somewhat different meaning from today's currently used definition in the brain mapping literature (Friston, 2011).

As discussed elsewhere in this volume (Baccalá & Sameshima, 2017), pairwise correlation, and its frequency domain equivalent, spectral coherence analysis, are limited in the information they provide due to their inherent symmetry, preventing detection of both interaction directionality and the presence of feedback. These limitations become even more important when one considers many more structures in their simultaneous interactions.

3. Adoption of causality concept in connectivity analysis

In 1969 Clive Granger introduced an idea of causality in econometric time series analysis. He made it precise by hypothesizing that if the use of the past information of a time series $x(n)$ significantly improves one's ability to predict another time series $y(n)$, then since cause must precede its consequence, one may say that $x(n)$ *Granger-causes* $y(n)$. The adjective 'Granger' is used to distinguish it from other uses of the word *causality* in both philosophy and system theory. This notion has been used in many flavors to qualify if one neural structure drives another, in a sense that the activity of a driving neural structure must precede that of one under influence (Baccalá & Sameshima, 2014b).

Since then, many related representations have flourished in both time and frequency domains, as summarized in Fig. 6.2. As can be seen on the left-hand side of this diagram, several authors have contributed to the time domain causality developments on pairwise time series analysis (Sims, 1972; Geweke, 1979, 1982, 1984), $N = 2$, which was then led with proper operational procedures to the implementation of dependency measurements, extending the Granger-causality concept to multiple time series analysis, $N > 2$. In 1993, Helmut Lütkepohl proposed a null hypothesis inference for the multivariate case of the Granger-causality test (see also Lütkepohl, 2005).

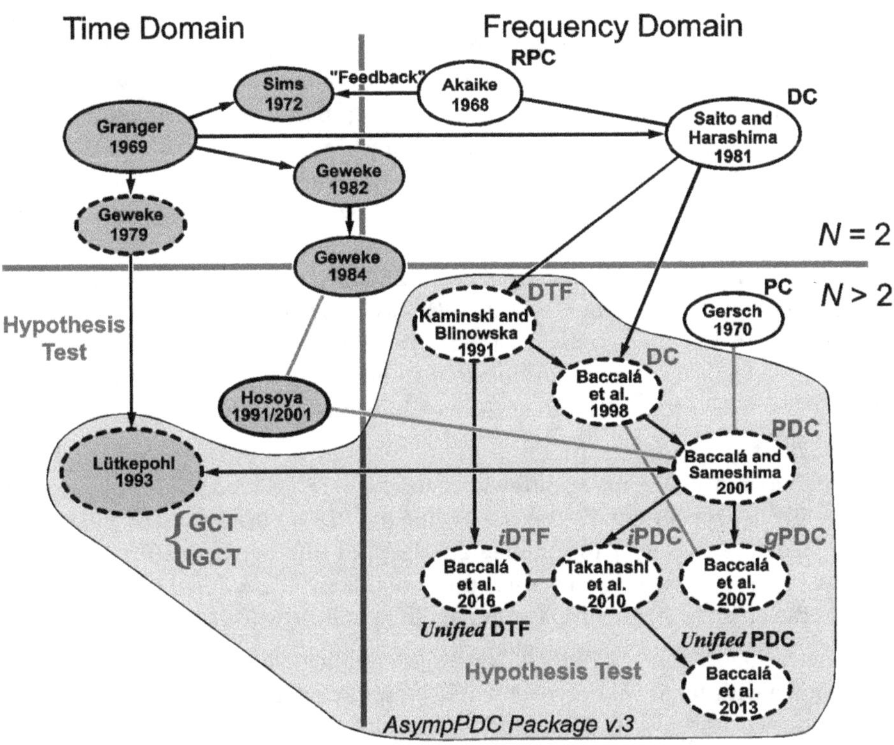

Figure 6.2: Diagram depicting the Granger-causality concept related development 'timelines', in which arrows and lines portray respectively the *dependence* and *similarity* between these contributions, without precisely indicating whether authors were fully aware of one another's works at the time of their publication. On the left-hand side the diagram shows time domain related causality concepts, while the frequency domain developments are shown on the right-hand side. Two important contributions by Geweke (1982, 1984) are depicted on the interface between time and frequency domain, in which the 1984 contribution also extended the analysis framework toward the multivariate case (N > 2). GCT: Granger causality test; IGCT: instantaneous Granger causality test; RPC: relative power contribution; DC: directed coherence; DTF: directed transfer function; PC: partial coherence; PDC: partial directed coherence; iDTF: information directed transfer function; gPDC: generalized partial directed coherence; iPDC information partial directed coherence. (Modified from Baccalá & Sameshima, 2014a, p. 7)

An early frequency domain contribution (Fig 6.2, right-hand side) came from Akaike (1968), conceiving the feedback possibilities with the measure called 'relative power contribution' (RPC), preceding Sims (1972), who reexplored Granger's (1969) proposal considering the notion of feedback detection. Note that Akaike's (1968) frequency domain approach also slightly preceded the publication of the landmark Granger (1969) time domain causality concept.

Saito & Harashima (1981) developed a frequency domain pairwise description applied to EEG analysis, called 'directed coherence' (DC), which was extended to the multivariate case by Baccalá et al. (1998), following the notion of directed transfer function (DTF) proposed by Kamiński & Blinowska (1991). These authors used multivariate autoregressive (MAR) modeling to fit the linear relationship of multiple time series. Inspired by Saito & Harashima's (1981) work, in 2001 we proposed a description using directly the multivariate autoregressive matrix coefficients, called 'partial directed coherence' (PDC), thus being more closely related to the concept of Granger-causality, and mirroring the Granger multivariate time domain causality test (GCT) (Lütkepohl, 1993; Baccalá & Sameshima, 2001). In other words, PDC is a frequency domain measure homologous to the time domain Granger causality test (see Sameshima et al., 2015), i.e. it 'measures' directed adjacent interaction, while DTF uncovers the reachability – of direct or indirect flow of information between pair of neural structures.

All three forms of DTF using different metrics – DTF, DC and iDTF – expose the reachability aspect of neural structures, while the PDCs – original PDC, gPDC and iPDC – are able to detect the immediate directed link between structures. Our experience shows that PDC describes the actual active directional links and DTF exposes the effect of signal propagation in the neural network.

We have made the *AsympPDC Package*[1] publicly and freely available, in consonance with the ideal of reproducible research, so that more researchers can use, test and improve it. The package contains the implementation of all measures shown within the shaded area in Fig. 6.2, by providing the respective null hypothesis tests based on rigorous asymptotic statistical procedures. Among the frequency domain descriptions our measures of choice are represented by what we call the information metric PDC (iPDC) and DTF (iDTF), because their magnitudes provide effect sizes and rigorous hypothesis test p-values, respectively for direct link and reachability strengths. For instance, a change in iPDC magnitude, when two controlled experimental conditions are compared, could mean physiologically that the actual connectivity strength of influence of those two neural structures has changed significantly through, for instance,

---

[1] AsympPDC Package v.3 can be obtained at http://www.lcs.poli.usp.br/~baccala/pdc/ .

neural plasticity mechanisms (Merzenich & Sameshima, 1993). By assessing connectivity detection based on multiple structures recorded and analyzed simultaneously, one could decrease the ambiguities that arise when only pairwise analysis (say cross-correlation or coherence) is performed.

There are many practical challenges for the application of connectivity inference methods, such as gauging the inadequacy of autoregressive model fitting, presence of outlier data contaminants, volume conduction issue in EEG recording, presence of nonlinear dynamics, slow and inadequate data analysis pipeline, etc., just to list few.

Even assuming that the latter technical limitations have been overcome and the true dynamic connectivity between neural structures lies revealed before us, in whatever guise, we must ask what help these connectivity descriptions give us in answering Arbib's first question.

4. Challenges

To address this question and what we face ahead, one must realize that despite their descriptive power, any representation of the relationship between neural structures, as has become fashionable through 'graphs', fancifully named the brain 'connectome' (Sporns, 2012), actually adds another layer with a world of new possibilities but also of complexities, where the *curse of dimensionality* rules and may make understanding even more difficult. There is no clear physiologically grounded and unifying brain theory proposal as to how one might benefit from these descriptions towards understanding how the brain processes information, generates behavior or performs cognitive function. There are many currently popular graph theoretical analysis proposals involving brain connectivity maps, in which concepts such as "small-world network", random graph, and other crafted measures have been extensively explored, but they lack down to earth interpretations for successful diagnosis and brain dynamic states detection (Sporns, 2011).

Perhaps today the core question is whether 'Big Data'-like approaches based on connectivity descriptions can prove to be useful. As with the traditional approach of purposeful experimentation from the dawn of neuroscience, it is clear to us that the carefully designed behavioral protocols are to remain the most important study tool, in which connectivity analysis techniques can help with inspiring good hypotheses that ultimately must go through validation via the same system intervention methods of early neuroscience experimentation, using adequate brain activity measurement technology by assessing both appropriate brain spatial scale and behaviorally related temporal scale, as depicted in Fig. 6.1. One realizes that 'the time is of the essence' in brain workings. As such, note that only electrophysiological techniques (EEG, ERP, MEG and ECo) are able to grasp and dissect neuronal and neural activities on the time scale of cognitive functions, i.e. on the hundreds of milliseconds span. This is where Granger causality based frequency domain connectivity techniques such as iPDC and iDTF could show their values by uncovering the direction of information flow, its magnitude, and the frequency bands of interaction.

Ultimately, we think that larger scale clinical/experimental trials to validate analysis techniques for each type of connectivity description are required. Reliably determining the where and the how when it comes to the brain remains a challenging problem that requires multidisciplinary approaches to integrate and validate data from different sources.

Bibliography

Akaike, H. (1968). On the use of a linear model for the identification of feedback systems. *Annals of the Institute of Statistical Mathematics*, 20, 425-440.

Arbib, M.A. (ed.) (1995). *The handbook of brain theory and neural networks*. Cambridge (MA): MIT Press.

Aertsen, A.M.H.J. & Gerstein G.L. (1991). Dynamic aspects of neuronal cooperativity: fast stimulus-induced modulations of effective connectivity. In: Krüger, J. (ed.). *Neuronal cooperativity*. Berlin: Springer, pp. 52-67.

Baccalá, L.A.; de Brito, C.S.N.; Takahashi, D.Y. & Sameshima, K. (2013). Unified asymptotic theory for all partial directed coherence forms, *Philosophical Transactions of the Royal Society A*, 371, article 20120158, 14 pp.

Baccalá, L.A. & Sameshima, K. (2001). Partial directed coherence: a new concept in neural structure determination. *Biological Cybernetics*, 84(6), 463-474.

——— (2014a). Brain connectivity: an overview. In: Sameshima, K. & Baccalá, L.A. (eds.). *Methods in brain connectivity inference through multivariate time series analysis*. Boca Raton (FL): CRC Press, pp. 1-9.

——— (2014b). Multivariate time-series brain connectivity: a sum-up. In: Sameshima, K. & Baccalá, L.A. (eds.). *Methods in brain connectivity inference through multivariate time series analysis*. Boca Raton (FL): CRC Press, pp. 245-251.

——— (2017). A neural link centered approach to brain connectivity. In: Adams, F.; Pessoa Jr., O. & Kogler Jr., J.E. (eds.). *Cognitive science: recent advances and recurring problems*. Wilmington (DE): Vernon Press (present volume), pp. 101-112.

Baccalá, L.A.; Sameshima, K.; Ballester, G.; Do Valle, A.C. & Timo-Iaria, C. (1998). Studying the interaction between brain structures via directed coherence and Granger causality. *Applied Signal Processing*, 5, 40-48.

Baccalá, L.A.; Takahashi, D.Y. & Sameshima, K. (2016). Directed transfer function: unified asymptotic theory and some of its implications. *IEEE Transactions on Biomedical Engineering*, 63, 2450-2460.

Friston, K.J. (2002). Beyond phrenology: what can neuroimaging tell us about distributed circuitry? *Annual Review of Neuroscience*, 25, 221-250.

——— (2011). Functional and effective connectivity: a review. *Brain Connectivity*, 1, 13-36.

Gersch, W. (1970). Epileptic focus location: spectral analysis method. *Science*, 169, 701-702.

Gerstein, G.L. & Perkel, D.H. (1969). Simultaneously recorded trains of action potentials: analysis and functional interpretation. *Science*, 164, 828-830.

——— (1972). Mutual temporal relationships among neuronal spike trains. *Biophysical Journal*, 12, 453-473.

Geweke, J. (1979). Testing the exogeneity specification in the complete dynamic simultaneous equation model. *Journal of Econometrics*, 7, 163-185.

——— (1982). Measurement of linear dependence and feedback between multiple time series. *Journal of American Statistical Association*, 77, 304-324.

——— (1984). Measures of conditional linear dependence and feedback between time series. *Journal of American Statistical Association*, 79, 907-915.

Granger, C.W.J. (1969). Investigating causal relations by econometric models and cross-spectral methods. *Econometrica*, 37, 424-438.

Gray, C.M.; Maldonado, P.E.; Wilson, M. & McNaughton, B. (1995). Tetrodes markedly improve the reliability and yield of multiple single-unit isolation from multi-unit recordings in cat striate cortex. *Journal of Neuroscience Methods*, 63, 43-54.

Hardy, G.H. (1940). *A mathematician's apology.* Cambridge (GB): Cambridge University Press.

Hosoya, Y. (1991). The decomposition and measurement of the interdependency between second-order stationary processes. *Probability Theory and Related Fields*, 88, 429–444.

——— (2001). Elimination of third-series effect and defining partial measures of causality. *Journal of Time Series Analysis*, 22, 537–554.

Kamiński, M. & Blinowska, K.J. (1991). A new method of the description of the information flow in brain structures. *Biological Cybernetics*, 65, 203-210.

Kwong, K.K.; Belliveau, J.W.; Chesler, D.A.; Goldberg, I.E.; Weisskoff, R.M.; Poncelet, B.P.; Kennedy, D.N.; Hoppel, B.E.; Cohen, M.S.; Turner, R.; Cheng, H.; Brady, T.J. & Rosen, B.R. (1992). Dynamic magnetic resonance imaging of human brain activity during primary sensory stimulation. *Proceedings of the National Academy of Science USA*, 89, 5951-5955.

Lütkepohl, H. (1993). *Introduction to multiple time series analysis.* New York: Springer.

——— (2005). *New introduction to multiple time series analysis.* New York: Springer.

McNaughton, B.; O'Keefe, J. & Barnes, C.A. (1983). The stereotrode: A new technique for simultaneous isolation of several single units in the central nervous system from multiple unit records. *Journal of Neuroscience Methods*, 8, 391-397.

Merzenich, M.M. & Sameshima, K. (1993). Cortical plasticity and memory. *Current Opinion in Neurobiology*, 3, 187-196.

Ogawa, S.; Lee, T.M.; Kay, A.R. & Tank, D.W. (1990). Brain magnetic resonance imaging with contrast dependent on blood oxygenation. *Proceedings of the National Academy of Science USA*, 87, 9868-9872.

Posner, M.I. & Raichle, M.E. (1994). *Images of mind.* Scientific American Library. San Francisco: W.H. Freeman.

Ramachandran, V.S. & Hirstein, W. (1998). The perception of phantom limbs. *Brain,* 121, 1603-1630.

Saito, Y. & Harashima, H. (1981). Tracking of information within multichannel EEG record: causal analysis in EEG. In: Yamaguchi, N. & Fujisawa, K. (eds.). *Recent advances in EEG and EMG data processing.* Amsterdam: Elsevier, pp.133-146

Sameshima, K.; Takahashi, D.Y. & Baccalá, L.A. (2015). On the statistical performance of Granger-causal connectivity estimators. *Brain Informatics,* 2, 119-133.

Sims, C.A. (1972). Money, income, and causality. *American Economy Review,* 62, 540-552.

Sporns, O. (2011). *Networks of the brain,* Cambridge (MA): MIT Press.

——— (2012). *Discovering the human connectome.* Cambridge (MA): MIT Press.

Takahashi, D.Y.; Baccala´, L.A. & Sameshima, K. (2010). Information theoretic interpretation of frequency domain connectivity measures. *Biological Cybernetics,* 103, 463–469.

Chapter 7

A neural link centered approach to brain connectivity

Luiz A. Baccalá
Polytechnic School
University of São Paulo (USP), São Paulo, SP, Brazil
baccala@lcs.poli.usp.br

Koichi Sameshima
Medical School
University of São Paulo (USP), São Paulo, SP, Brazil
ksameshi@usp.br

Abstract

The last couple of decades has seen a methodological shift in the characterization of brain dynamics, as the early prevalent search for areas of increased neural activity under stimulus, often termed 'neophrenology', is slowly being replaced by investigations that focus on how those areas interact. The general term 'connectivity' has been used far and wide to describe this trend. Early on, however, the need was felt for qualifying what one means by connectivity with more precision. Adjectives like 'effective' and 'functional' connectivity are in common use today, even though their application is often inconsistent throughout the literature. This fuzzy scenario has been recently compounded by the advent of Granger causality based connectivity estimators that expose the directionality of information flow in addition to the presence of mere correlation, leading to the need for establishing a more general and consistent classification of how one should interpret connectivity issues, which we address by touring the historical evolution of these ideas.

1. Introduction

Not so long ago, mostly due to technological limitations, neurophysiological investigation was performed under what one might call a 'pairwise' paradigm: stimulate, or observe the activity at one structure, and then look for what is going on at other structures, one at a time. If matching activity is found, one may speak of the presence of connectivity between the original structure and the 'responding' one.

Decades of mapping effort went by and coincided with the evolution of technology towards the simultaneous measurement of the neural activity at many neural structures, via multi-electrode electroencephalography (EEG) (in the order of hundreds of electrodes) and imaging – most specifically functional magnetic resonance imaging (fMRI), which captured the imagination of scientists and laymen alike, as one now could 'see' what brain areas become active as subjects exhibit behavior – landing us, with the popular media metaphor of areas alight in the brain under this or that circumstance, into a picture that is reminiscent of 19th century phrenology (Uttal, 2001).

Exciting and aesthetically appealing as this surely is, came the slow realization that beautiful pictures are not enough, and that to understand how the brain drives behavior one must also investigate how the lit-up areas interact.

Early on, to do so, the first idea was to look for the relationship between areas of significant activity via the time tested idea of *correlation* due to Pearson and Galton (Stigler, 1989), by now more than a century old. The concept has become so culturally embedded and familiar as to lead to an age derived comfort and widespread use that has fostered dangerous thoughtless abuse.

When comparing waveforms, the correlation coefficient measures shape similarity (Fig. 7.1a) and if relative time displacement is allowed, it gives an idea of latency embodied by cross-correlation functions – where a correlation coefficient for each relative waveform delay is computed (Fig. 7.1b). An assumption is that activity wave forms do not become too distorted as they propagate. Thus, high correlation implies that a connection cannot be ruled out.

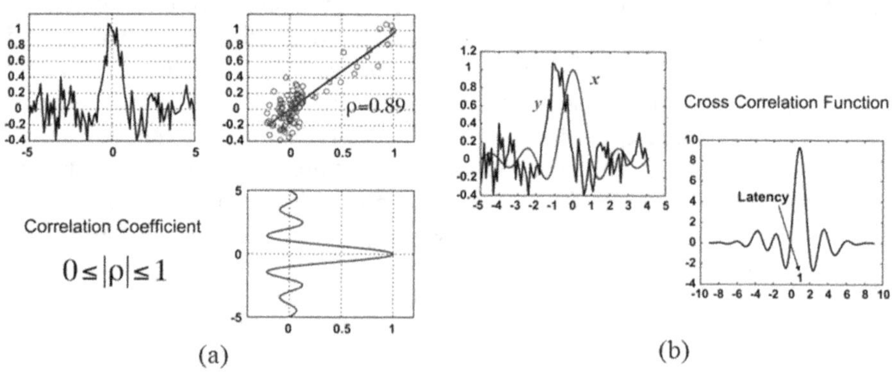

Figure 7.1: (a) Comparison between wave shapes with mutual time synchronous regression revealing approximate shape similarity as ρ is high. (b) Cross-correlation function between waveforms depicting a latency of 1 time sample. (All figures prepared by the authors.)

Specially in EEG analysis, because cross-correlation functions have a natural frequency domain representation – the cross-spectrum – it is possible to consider what happens in typical EEG frequency bands whose ranges have a well known associated physiological interpretation (Başar, 2004). After proper normalization, cross-spectra lead to the notion of spectral coherence (Bendat & Piersol, 2010), which is no more than the correlation coefficient computed on a frequency band basis.

Thanks to its familiarity and the comfort its age brings, correlation has been the basis of a multimillion dollar initiative to draw the human 'Connectome' – in an allusion to the 'Genome' (Sporns, 2011; Seung, 2012).

The problem is that strict correlation analysis is of limited value (see Sec. 2) and must be dismissed in favor of more general characterizations (Secs. 3 and 4), which imply a novel interpretation that departs from a response activity centered view of the brain towards a more link/interaction centered brain characterization (Sec. 5).

2. The limits of correlation

Ironically, despite its classic appeal, correlation analysis has a number of drawbacks. Correlation coefficients and frequency domain coherence are symmetric: they only tell whether the activities of two structures are tied via the similarity of their wave shapes, and more importantly, it is impossible to state whether either one dominates the other or whether some kind of feedback between them is present.

Perhaps more importantly, descriptive incompleteness is compounded when many structures are measured simultaneously. In fact, as far back as 2001, we employed a toy dynamic model with its 'connectivity' imposed by design and investigated the use of pairwise cross-correlation function analysis to 'reverse' engineer the structure (Baccalá & Sameshima, 2001a). What we found is that not only ambiguities arise, but also that inference errors are inevitable.

If simple cross-correlation analysis has shortcomings, is there an alternative? The answer is yes, but we must move on to consider the full machinery of time series analysis.

3. Granger causality

Before considering the general multivariate case, let us first examine the case of pairs of time series. The history of the subject is long, but we focus on the contribution of Granger (1969) and his concern with how interest rate and inflation relate: is one responsible for the other? All he had were observations over long periods of time. Questions like this amount to something like the egg-chicken conundrum: which one came first? He proposed the idea that if one could observe changes in one quantity that helped predict changes in the other quantity, one could make the case for some sort of 'causality', whereby connectivity by means of physical influence is a possibility – as through neural projections able to carry the message, in our case. In fact, the term *causality* is a marketing miracle, which, much as with the term love, can lead to many rather subtly different interpretations. In many areas, like philosophy (Coventry, 2006), system theory and physics (Nussenzveig, 1972), and probability theory (Pearl, 2000), it has acquired many distinct technical nuances. Here Granger's proposal is referred to as G-causality (or GC), to distinguish it from the latter interpretations; its non-interventional character sets it apart, since it just uses 'consistent prediction' improvement as a criterion for its presence. In fact, one may detect prediction improvement, i.e. GC, even when no actual physical linking mechanism exists. To understand this, remember that weather reporters, despite some predictive forecast capability, are unable to make actual rain happen.

In practice, GC is inferred by inspecting the statistical significance of the nullity of coefficients of a regression of the present of one series $x(n)$ based on the past of another series $y(n)$. GC of $x(n)$ onto $y(n)$ cannot be rejected if $x(n)$'s past improves $y(n)$ predictability.

This can be accomplished by regressing the $y(n)$ series on the past of $x(n)$ in addition to its own past:

$$y(n) = \sum_{k>0} a_{yy}(k)\, y(n-k) + \sum_{k>0} a_{yx}(k)\, x(n-k) + e_y(n) \tag{7.1}$$

where significant GC cannot be rejected if $a_{yx}(k) \neq 0$ for some lag k.

By reversing the argument, if $a_{xy}(k) \neq 0$, one can investigate whether $y(n)$ G-causes $x(n)$. Since there is no mandatory reason for $a_{yx}(k) \neq 0$ to imply $a_{xy}(k) \neq 0$, one immediately sees that, unlike correlation, G-causality is not symmetric. It is possible for $x(n)$ to G-cause $y(n)$ without $y(n)$ G-causing $x(n)$.

This allows placing influence arrows between time series. Thus, as opposed to correlation based notions whose relations are symmetric and reciprocal, GC-based connectivity can, by contrast, be described as a directional quantity that measures the direction of information flow and the eventual presence of feedback. Therefore, time series analysis based GC, even if just pairwise, already provides a level of description that supersedes correlation analysis.

3.1 Pairwise frequency domain description

Again, because of the importance of frequency domain descriptions, specially in EEG, an early rendition of GC appeared in the work of Saito & Harashima (1981) and was named 'directed coherence'. It only achieved a more definitive form in the pairwise paradigm proposed in 1985 under the name of 'Directed Transfer Function' (DTF) (Franaszczuk et al., 1985).

The core rationale comes from a canonical way of writing the spectra of a pair of time series in terms of filtering white noise (a.k.a. innovations noise), which is dynamically rich and contains equal amounts of power at all frequencies. Under this representation, the dynamical relationships between the observed time series are encoded with the help of suitably defined filters $H_{ij}(f)$ (see Fig. 7.2) that sift the dynamics of interest from amid the innovations noise. In this representation, one may show that $H_{yx}(f) = 0$ translates into lack of directed influence from $x(n)$ onto $y(n)$. The term 'directed coherence' is due the latter quantity's appearance in the factorization of the coherence between time series (Baccalá & Sameshima, 2001b). Unsurprisingly $a_{yx}(k)$ nullity implies $H_{yx}(f)$ nullity when one is considering just a pair of time series.

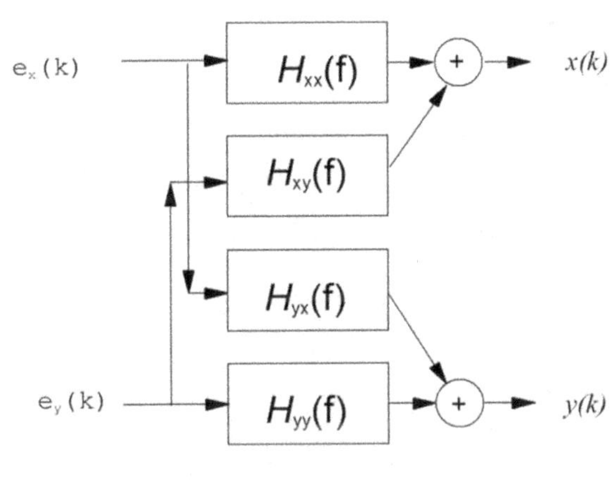

Figure 7.2: White noise filtering representation of time series. If $H_{yx}(f) = 0$ for all f, then $x(n)$ does not GC $y(n)$.

4. The multivariate case

Close examination of pairwise time series analysis in the spirit of Baccalá & Sameshima (2001a) easily shows that GC/DTF are limited when it comes to reverse engineering large networks of dynamically interconnected simultaneously observed variables, even though it allows directional information flow arrow and measurement between them.

Since time series representations are not restricted to simultaneously observed pairs, Kamiński & Blinowska (1991) used a multivariate (a.k.a. vector) autoregressive (MAR or VAR) modeling (Lütkepohl, 2005) to write the joint power spectral matrix density whose factorization generalizes the filtering scheme of Fig. 7.2 to multiple time series. This appeared to provide a definitive solution to the problem until we noticed (Baccalá et al., 1998) that structural connectivity predictions based on DTF did not match multivariate generalizations of Granger causality (Lütkepohl, 2005). The reason is the indirect dependence of DTF on the $a_{ij}(k)$ model regression coefficients in the multivariate case.

To remedy this, we proposed the notion of 'partial directed coherence', PDC for short (Baccalá & Sameshima, 2001b), which can be shown to be a factorization of partial coherence (Takahashi et al., 2010). PDC is identically zero if $a_{ij}(k) = 0$ for all k, so that it closely matches canonical multivariate GC tests in the literature (Lütkepohl, 2005). The distinction between DTF and PDC is best appreciated by a simulated example whose time series $x_i(n)$, $(1 \leq i \leq 5)$, have their dynamics described by:

$$x_1(n) = -0.5\, x_1(n-1) + \sqrt{2}w_1(t) \tag{7.2}$$

$$x_2(n) = -0.5\, x_2(n-1) + y(n-1) + 3.53w_1(n) + 2.74w_2(n) + 0.65x_5(n-4) \tag{7.3}$$

$$x_3(n) = 0.5\, x_2(n-1) + w_3(n) \tag{7.4}$$

$$x_4(n) = +G\, x_2(n-2) - 0.5x_3(n-1) + 0.25x_4(n-1) + w_4(n) \tag{7.5}$$

$$x_5(n) = 2 \cdot 0.95 \cdot \cos(3\pi/2)x_5(n-1) - 0.95^2 x_5(n-2) + w_5(n) \tag{7.6}$$

$$y(n) = 0.97\sqrt{2}y(n-1) - 0.97^2 y(n-2) + x_1(n-1) \tag{7.7}$$

Eq. 7.7 represents the unobserved variable $y(n)$ added to account for a filtering operation that takes place in the link from x_1 to x_2. This link filtering is highly frequency selective by design. Also note that the letter G in Eq. 7.5 was singled out for illustrative purposes. The $w_i(n)$ quantities stand for white mutually uncorrelated innovations noise.

From a connectivity standpoint one might diagrammatically sum up the equations in Fig. 7.3, where emphasis was given to the filtered x_1 to x_2 link by using a rectangle in the arrow.

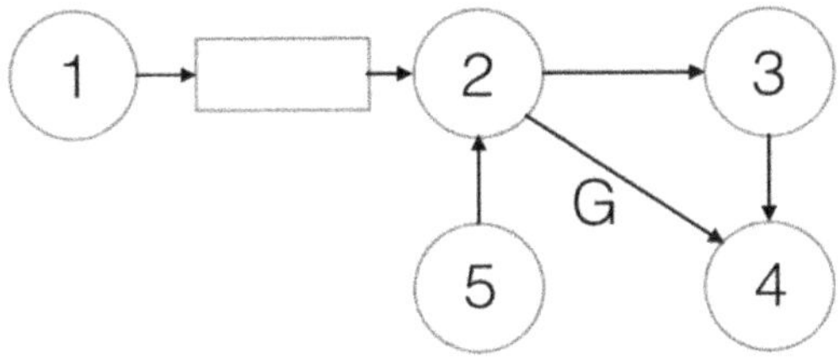

Figure 7.3: Graph representation of the connectivity example from the text, where the rectangle represents the presence of the connection filter (Eq. 7.7).

In Fig. 7.4a, $n_s = 5000$ data points are used to estimate a VAR model (maximum past lag $p = 9$) which shows iPDC, an information theoretic normalization based PDC (Takahashi et al., 2010). The panel of graphs is presented in the so-called *standard form*, where signal 'sources' correspond to columns and signal 'sinks' to rows. Power spectral densities (with gray backgrounds) for each series are depicted along panel diagonal graphs, in arbitrary units. The reader will notice the one-to-one mapping of significant iPDC values for existing connections in Fig. 7.3. The oscillatory behavior of x_2 stands out as due to the resonant filter interaction represented by Eq. 7.7 operating on signal x_1.

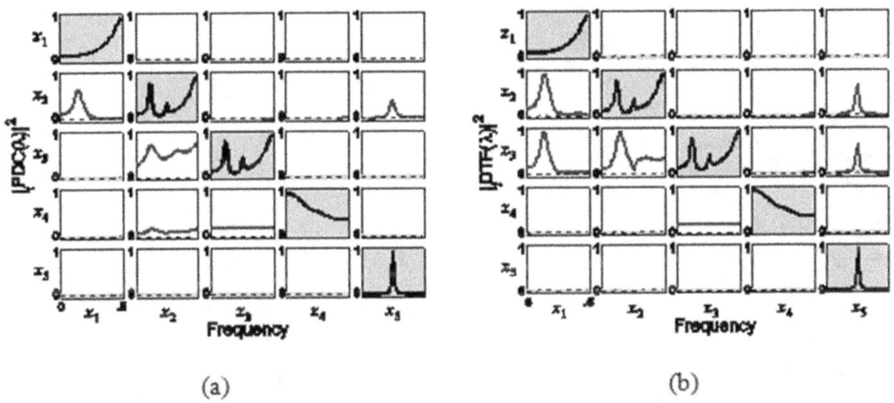

Figure 7.4: Standard form representations (see text) for (a) iPDC showing one to one correspondence with Fig. 7.3 connectivity and (b) iDTF with G = 0.2438 in the text example. Grayed background graphs in the standard form panel refer to time series power spectra. Most importantly note the absence of significant expected DTF connectivity from x_2 to x_4 as well as x_5 to x_4, since physical pathways exist which are nonetheless rendered *inactive* in this example from G's value choice. Dashed lines represent significance thresholds. Connectivity is to be read from column to row.

By contrast, information DTF (iDTF) (Takahashi et al., 2010) in Fig. 7.4 differs from iPDC by being large only for connections which represent active pathways from one node to another. Such is the case for the paths from x_1 to x_3 and x_5 to x_3, both via x_2. Note that even though there are direct immediate links from x_2 to x_3 and to x_4, the crafty choice of G = 0.2438 leads to signal cancellation between the latter pathways, so that the overall influence from x_2 to x_4 is negligible, i.e. neither signals x_1 nor x_5, that go through x_2, nor x_2's signal can reach x_4. Only the influence originating locally in x_3 reaches x_4.

These observations show that whereas PDC describes the actual active directional links, DTF captures the effect (and also eventual blockage) of signal propagation within the network – i.e. the existence of active signal pathways.

This example also has another pedagogically convenient trait, when the PDC relationship between x_1 and x_2 is contrasted with results from coherence analysis (Fig. 7.5), where one readily sees the dip in coherence (gray lines) around one of the oscillatory frequencies in x_2, as opposed to PDC's clear large effect. This confirms the importance of x_1 onto x_2, even as a generator of that oscillatory feature by means of the characteristics of the directed (resonant) link that joins them.

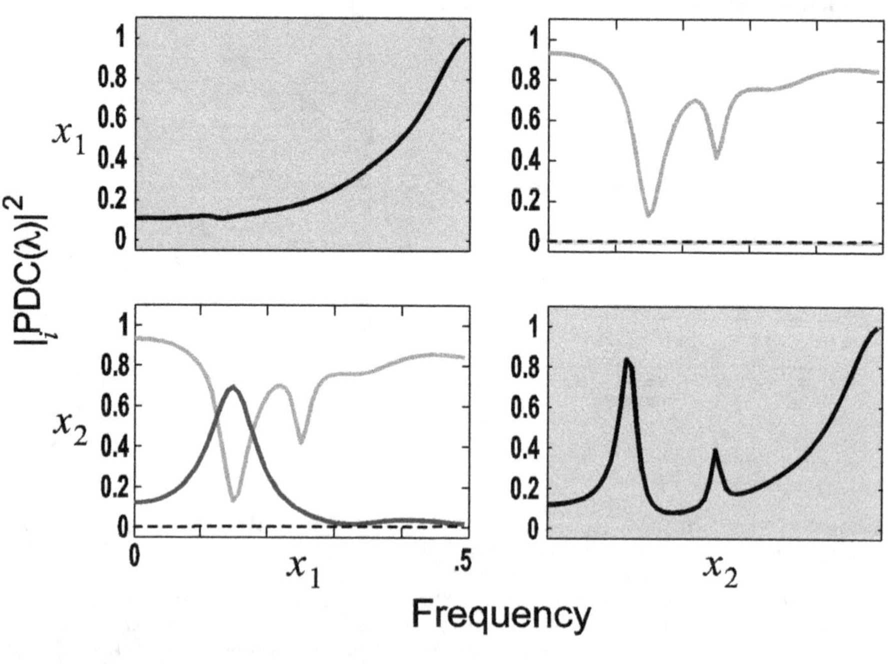

Figure 7.5: iPDC detail from Fig. 7.4a, with superposed coherence (gray) which decreases at the very frequency where the influence from x_1 to x_2 is largest.

Thus, close examination of the above differences shows that PDC captures the immediate directed link structure, whereas DTF exposes 'signal' reachability, i.e. mutual influence via signals that may have to travel through intermediate structures before effects are felt, and which may even cancel as in the case of the influence of x_2 onto x_4. Thus, DTF and PDC are complementary connectivity characterizations and the often raised issue as to which measure is better is empty. Their joint description adds to our understanding of the dynamical interaction between series.

5. Discussion and implications

The systematic use of time series tools provides a sort of richness beyond that of simple correlation/coherence analysis, and it is quite clear that 'Connectome' efforts that overlook this are unnecessarily limited. In fact, this very richness demands that the way to look at connectivity be reconsidered towards a link centered paradigm, where the detailed description of links is at least as important as that of regions that display physiologically relevant activity, as becomes undeniably clear when the two very distinct dynamics of x_1 and x_2, and thus of little mutual coherence, are nonetheless strongly linked due to how the link is shaped.

Due to their ubiquity, the hope of fully replacing the widespread notions of effective/functional connectivity is somewhat unrealistic. Yet, how one sees and considers connectivity is in need of an upgrade, because knowledge of whether and how a physiological link is active at one given instance is important, being crucial whether this link is between immediate structures, as exposed by PDC, or via the propagation through multiple pathways, as revealed by DTF. This naturally leads to the classification of directedness in terms of being active/inactive via immediate links or via indirect pathways (Baccalá & Sameshima, 2014).

What marks this proposal as different from effective connectivity is the latter's reliance on 'a priori' model driven considerations. The notion of functional connectivity is even weaker, as it is tied to correlation based considerations (Friston, 1994).

Time series analytical methods via PDC/DTF constitute a data-driven alternative unhindered by constructive model preconceptions, whereby hypotheses for experimentation may be generated.

The attractiveness of the methods discussed here is compounded by the fact that PDC (Baccalá et al., 2013) and DTF (Baccalá et al., 2016) have rigorously derived statistics that serve the purpose of objective hypothesis testing.

One current challenge to the widespread use of techniques such as the ones discussed here is their relatively recent introduction (compared to correlation ideas), which can only be overcome with time, as users begin to delve deeper in the arts of time series analysis, whose merits and technical limitations must be clearly understood.

Many issues have been left untouched here, whilst some were only glossed over briefly; the interested reader may find additional material, that includes further discussion of the historical perspective on this theme, in Sameshima & Baccalá (2014).[1]

[1] The reader may find the model behind the illustrations in http://www.lcs.poli.usp.br/~baccala/pdc/ebiic_paper.html, where it is possible to change the parameter G to open gates that close the x2 to x4 interaction. This is done by using a script under the AsympPDC time series package, available from http://www.lcs.poli.usp.br/~baccala/pdc/asymp_package_v3.zip.

Bibliography

Baccalá, L.A.; de Brito, C.S.N.; Takahashi, D.Y. & Sameshima, K. (2013). Unified asymptotic theory for all partial directed coherence forms. *Philosophical Transactions of the Royal Society A*, 371, 20120158, 1-14.

Baccalá, L.A. & Sameshima, K. (2001a). Overcoming the limitations of correlation analysis for many simultaneously processed neural structures. In: Nicolelis, M.A.L. (ed.). *Advances in neural population coding*. Amsterdam: Elsevier. *Progress in Brain Research*, 130, 33-47.

——— (2001b). Partial directed coherence: a new concept in neural structure determination. *Biological Cybernetics*, 84(6), 463-474.

——— (2014). Causality and influentiability: the need for distinct neural connectivity concepts. In: Ślęzak, D.; Tan, A.-H.; Peters, J.F. & Schwabe, L. (eds.). *Brain informatics and health*. Cham (CH): Springer. *Lecture Notes in Artificial Intelligence*, 8609, 424-435.

Baccalá, L.A.; Takahashi, D.Y. & Sameshima, K. (2016). Directed transfer function: unified asymptotic theory and some of its implications. *IEEE Transactions on Biomedical Engineering*, 63(12), 2450-2460.

Baccalá, L.A.; Sameshima, K.; Ballester, G.; Do Valle, A.C. & Timo-Iaria, C. (1998). Studying the interaction between brain structures via directed coherence and Granger causality. *Applied Signal Processing*, 5, 40-48.

Başar, E. (2004). *Memory and brain dynamics: oscillations integrating attention, perception, learning, and memory*. Boca Raton (FL): CRC Press.

Bendat, J.S. & Piersol, A.G. (2010). *Random data: analysis and measurement procedures*. 4th ed. New York: Wiley.

Coventry, Angela M. (2006). *Hume's theory of causation*. New York: Bloomsbury.

Franaszczuk, P.J.; Blinowska, K.J. & Kowalczyk, M. (1985). The application of parametric multichannel spectral estimates in the study of electrical brain activity. *Biological Cybernetics*, 51, 239-247.

Friston, K.J. (1994). Functional and effective connectivity in neuroimaging: a synthesis. *Human Brain Mapping*, 2, 56-78.

Granger, C.W.J. (1969). Investigating causal relations by econometric models and cross-spectral methods. *Econometrica*, 37(3), 424-438.

Kamiński, M. & Blinowska, K.J. (1991). A new method of the description of the information flow in brain structures. *Biological Cybernetics*, 65(3), 203-210.

Lütkepohl, H. (2005). *New introduction to multiple time series analysis*. New York: Springer.

Nussenzveig, H.M. (1972). *Causality and dispersion relations*. New York: Academic.

Pearl, J. (2000). *Causality: models, reasoning, and inference*. Cambridge (GB): Cambridge University Press.

Saito, Y. & Harashima, H. (1981). Tracking of information within multichannel EEG record: causal analysis in EEG. In: Yamaguchi, N. & Fujisawa, K. (eds.). *Recent advances in EEG and EMG data processing*. Amsterdam: Elsevier, pp. 133-146.

Sameshima, K. & Baccalá, L.A. (eds.) (2014). *Methods in brain connectivity inference through multivariate time series analysis*. Boca Raton (FL): CRC Press.

Seung, S. (2012). *Connectome: how the brain's wiring makes us who we are*. Boston: Houghton Mifflin Harcourt.

Sporns, O. (2011). *Networks of the brain.* Cambridge (MA): MIT Press.

Stigler, S.M. (1989). Francis Galton's account of the invention of correlation. *Statistical Science,* 4(2), 73-79.

Takahashi, D.; Baccalá, L. & Sameshima, K. (2010). Information theoretic interpretation of frequency domain connectivity measures. *Biological Cybernetics,* 103, 463-469.

Uttal, W.R. (2001). *The new phrenology: the limits of localizing cognitive processes in the brain.* Cambridge (MA): MIT Press.

Chapter 8

Simulation and adaptivity

João José Neto
Polytechnic School
University of São Paulo (USP), São Paulo, SP, Brazil
jjneto@gmail.com

Abstract

This work briefly discusses and collects requirements for the development of simulators based on complex models, e.g. intended for imitating natural phenomena such as the behavior of living beings and mind. In order to achieve that, several concepts and techniques are chosen from different fields of interest, with no purpose of being exhaustive. The text published here is an attempt to establish some directions for the conception and construction of general event-driven simulators for complex reactive systems, described as rule-driven models, and implemented within the framework of adaptivity.

1. General issues

1.1 Introduction

The present work develops an introductory study on the subject Simulation and Adaptivity, from the simulation designer's viewpoint, towards the establishment of a framework for modeling, simulating and investigating properties of the objects of the simulation process. Obviously, there isn't enough space here for a thorough examination of all aspects of a project like this, therefore we chose to discuss only few among the numerous themes of interest, leaving the remaining topics for another occasion. Our approach is to bring together the most important concepts from these areas, and to briefly investigate how teams who work with both simulation and adaptivity can mutually complement their skills in order to achieve adequate goals of common interest.

In this direction, we focus on those issues that are required for conceiving and designing complex systems, and for obtaining physical implementations in some available digital computer. With such a software, it is possible to represent and run, in a usual computer, good approximations of extremely complex processes, described by nontrivial computational representations.

Such simulation procedures can be projected for, at least partially, trying to imitate highly intricate processes, such as those currently observed in the many branches of the natural sciences. In the present text, some of our analyses are illustrated with instances of certain characteristics chosen among natural phenomena, such as attitudes found in biological and mental processes, and some features observed in complex natural phenomena.

1.2 Modeling

Modeling deals with computer representations of systems, objects and phenomena. Particularly interesting is the conception and design of models for particularly nontrivial systems, such as, for example, those related with nature, life and mind. Simulating complex phenomena usually calls for a good understanding of their operation, and considerable skill is required for correctly mapping them onto mathematical formulations that accurately model them, and can be implemented as computer programs that imitate their operation as precisely as possible.

The way such abstractions are created and constructed depends on their author's knowledge of the phenomenon to be simulated and on his/her creativity and mathematical abilities as well. Therefore, the suitability of a model or a formal statement adopted for representing the desired phenomenon is strongly influenced by how its authors regard objects populating their field of interest.

The simpler the target objects of the simulation, the easier will be modeling and simulating them, and (maybe) the closer to reality the results obtained through such simulation will be. However, highly sophisticated entities are seldom easy to model, and consequently their required underlying mathematical formulation, computer representation, modeling and simulation will likely become correspondingly challenging. In such a situation, simulation processes tend to become large and complex, and are also likely to become considerably time-, space- and resource-demanding.

One way to decrease costs, while still (partially) preserving the accuracy of expensive simulation processes, is to adequately moderate their resource demands. That effect may be achieved by reducing the level of detail represented by the models (e.g. by excluding complexities not so significant for the current application of the simulator, or by replacing detailed and complete complex analytical formulations by simpler and perhaps less accurate ones, etc.). Computer approximations of detailed and expensive models of complex phenomena are usually a practical way to make it feasible to study them in different (although less accurate) levels of representation. Such approach allows one even to carry out experimental methods on the computational objects, instead of on the actual phenomenon being studied.

Obviously, this approach usually cannot provide the most accurate results, but it tries to imitate cheaply, by means of approximate modeling, phenomena that are either too complex, hard to reach or difficult to probe directly, non-observable, non-controllable, too expensive, too small or too large, or too distant in space or in time. So, there are lots of phenomena, for example, many atomic and astronomical processes, that cannot be stimulated precisely, and some of them cannot even be observed directly, requiring indirect measurement and even pure theoretical speculation, which is the case of phenomena hypothetically occurring inside black holes, and of processes occurred too long ago, instants after the Big Bang.

In the absence of alternatives, simulation is usually an excellent alternative solution, working based on existing (when available) knowledge about the desired phenomenon, on theoretical hypotheses and on mathematical abstractions. In this way, simulation becomes a valuable tool that proves to be very effective for allowing practical experiments with the phenomenon being studied, even if the actual phenomenon is non-controllable or even non-observable. Many speculations performed this way have produced important insights for experimental research held in otherwise unapproachable fields, and have led to theories and models consistent with the current scientific knowledge.

Meanwhile, other researches have led to the improvement of existing models and even encouraged the discovery of yet unexplored facts on the subject being studied and its associated phenomena. When a desired phenomenon it is too difficult to model, we are usually compelled to avoid the difficulty by using some alternate model that approximates and simplifies the original one. So, resource-demanding phenomena require careful planning in order to choose acceptable simplifications and approximations that allow the construction of simpler, faster and feasible simulators.

Since this text is restricted to some considerations on modeling, simulation and adaptivity, important related topics such as human-computer interfaces will not be covered here, nor the important reasoning on the particular simulation technique to be chosen among the variety of mathematical models, the ways to perform simulation tasks and to implement simulation systems. For information on many important issues not covered here, we address the reader to the available literature on problem solving, modeling and simulation, for instance, Ross (2002), Chung (2004), and Hochbaum (1997).

1.3 Design choices

Among the techniques usually employed in the simulation of computing applications, *rule-based modeling* will be our choice here, since it is the approach that best and most naturally fits the requirements for the implementation of the self-modifying systems whose underlying device's operation is described by a set of rules (typically "if... then" rules).

In the scope of this work, three concepts are brought together: rule-based modeling, event-driven simulation and adaptivity. Such a fruitful combination provides a natural way of easily implementing discrete event simulators, designed to exhibit the desired self-modifying features. Discrete events and the event-driven simulation technique fit perfectly the requirements of a simulation scheme intended to represent and manipulate all information already obtained from the events that happened in the simulated system, while anticipating predictable events yet to occur, strictly on basis of known properties of the components of the system being simulated.

2. Simulation issues

2.1 Simulations and simulators

Computer programs that try to somehow reproduce mathematical, physical, biological, psychological or any other natural or artificial phenomena are often called "simulators" of such phenomena.

Self-modification (here referred to as "adaptivity") is not a new concept, since it has been extensively used to implement programs in early computers. However, for several reasons related to software management, involving the preservation of the system's and program's integrities, the emerging Software Engineering discipline of the 1970's almost banned such practice among programmers. For many years, experts discouraged building programs that change their own code, arguing that such a feature would make it significantly difficult to track down and evaluate their correctness, performance and behavior.

Presently, the situation isn't so radical anymore, since in several special-purpose applications the use of self-modification properties is being practiced again, in many particular application areas, in order to accomplish special effects that are difficult to achieve by other means.

This work is an attempt to bring together some issues on modeling and simulation, in order to open a discussion on the impact of exploring self-modifying features in such a context. The theme is very broad and there is no intention of exhausting the subject in this publication, so we will concentrate our attention only on a particular class of devices, the rule-driven devices, that turned out to fit nicely the implementation needs of self-modifying models.

2.1.1 Simulations

Simulation is a very popular computer technique, used in almost any field of knowledge, and is intended to replace, whenever possible, in a digital computer, actual objects with their corresponding software equivalent. This is not easy to achieve, since replacing actual objects imposes time constraints, functional requirements, and further restrictions that may be difficult to obtain or may even be unfeasible in available computers. However, by applying certain modeling simplifications, and maybe adding hardware features, simulators can be improved in order to react almost in real time to their stimulating events. Although simulators in general tend to operate slower than concrete phenomena, under favorable conditions some of them may perform well enough to actually replace the real phenomenon without important loss of performance.

In this way, simulators can almost convert a computer into a replica of the actual object under simulation (as far as their behavior goes), and the higher the required accuracy and level of detail, the higher will be their corresponding cost of utilization. A good practice when designing a simulator for use in such special applications is to try to balance detail, accuracy and performance levels.

In order to create good simulators, it is necessary to adequately model the target of simulator's attention. There are many ways to build models for the desired phenomena: we can model them analytically through equations, stochastically through statistical formulations, physically through the physical construction of analog (electrical, mechanical, pneumatic, hydraulic, etc.) models, or algorithmically, by means of computer programs, such as certain simulators that react to the simulated events.

The behavior associated to rule-driven devices may be encoded as a finite set of rules that are interpreted by an event-driven simulator. Such devices are very adequate for use in simulation processes, since they allow quite simple models that are powerful enough to represent very sophisticated phenomena. Generality and efficiency can be achieved through powerful, inexpensive, compact and effective simulators.

2.1.2 Event-driven engines

Event-driven engines are programs that can react to external stimuli (also called "events"). A usually good way to achieve practical simulation programs is to start by constructing event-driven models that rigorously describe the desired phenomenon.

The operation of event-driven engines resembles that of digital processor architectures with an interrupt system. Whenever an event happens to occur, the activation of a corresponding handling routine is scheduled, the execution of which takes place by performing whatever actions are needed for managing the event's occurrence. A series of such handlings, conveniently arranged, can track the sequence of actual events and may be programmed in order to adequately accomplish the chosen task.

This kind of (event-driven) engine provides therefore a very natural environment for the simulation of systems that respond to the occurrence of events (reactive systems), so that changes in the environment may be detected and asynchronous events may be captured by the system, and the corresponding handling routine can then be selected and executed. Event-driven simulation is achieved by means of simulators that can handle precedence orders and sporadically time. The chronology of occurrence of events is to some extent reflected by the ordering with which the system acknowledges events. Such ordering is also reflected in the execution sequence of the associated simulation actions.

There are two kinds of greater classes of events: independent events, which do not depend on actions taken during the simulation process, and dependent events, which originate within the models that perform the simulation procedures. In other words, independent events occur spontaneously and are usually received from the outside or arbitrarily generated in time; they can model attributes (such as delays) undertaken by components of the system during the simulation process. In the real world, independent events correspond to spontaneous physical occurrences, external to the phenomenon, while dependent events refer to reactions exhibited by its components.

One further class of events, named "impulsive events", is employed in some special situations in this work, and is very appropriate for simulating decisions that are taken on the basis of intuition, instinct or proactivity. Such phenomena do depend on history and on previously gathered knowledge, but they may occur independently of a strict logical combination of available information or knowledge, since sometimes they spontaneously emerge from one's will.

Although responses to impulsive events do not constitute authentic reactions to externally received stimuli, in the simulation scheme proposed here they can be reduced to equivalent reactive responses to spontaneously inserted artificial stimuli. Informally, in contrast to truly independent events, the sequence of which occurs "statically", before the simulation process is started, impulsive events may be considered as independent events that are scheduled dynamically (e.g., externally, at arbitrary instants in real time, or internally, arbitrarily at any instant of the simulation time). The following illustration (Fig. 8.1) sketches the operation of a typical event-driven simulator.

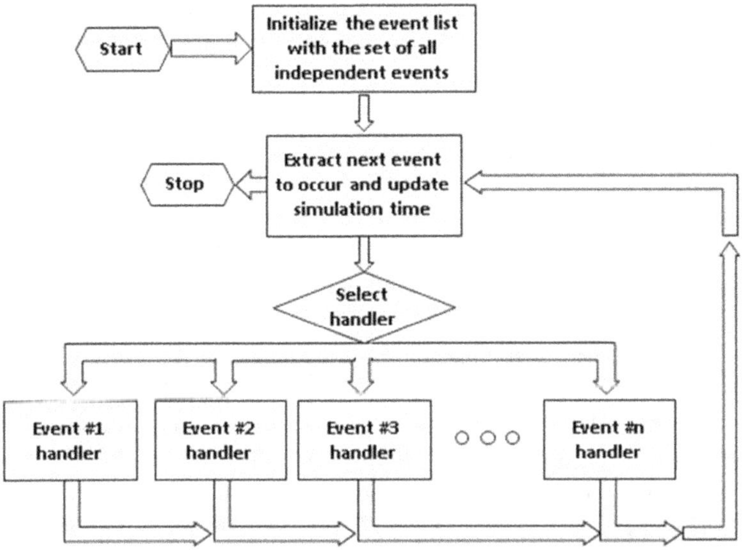

Figure 8.1: Event-driven simulator. (All figures prepared by the author.)

By first initializing externally all desired independent events, the event-driven simulator is configured to perform its next simulation according to the particular requirements imposed by the independent events supplied, and the simulator becomes satisfactorily set for starting its simulation process. In its main loop, the simulator extracts, for being interpreted and correspondingly processed, the first element of its event list (the element at the list head). Afterwards, the simulator checks the nature of the extracted event, and determines which of the available handling procedures specifically corresponds to that event. Finally, the specific handler associated to the chosen event is executed, in order to perform the simulation actions related to the occurrence of the event. The simulation loop is now able to resume by extracting and processing the first element of the remaining event list, and so forth, until no further events remain in the list, indicating that the simulation is over.

2.1.3 Behavior and its modeling

The previous general discussion on the simulation of processes indicates the importance of understanding well not only the particular phenomenon of interest, but the overall approach of realistically replacing it by some adequately chosen approximate model or its computational simulating counterpart. The previously mentioned processes, and their corresponding models and simulators, do not show dynamic features in their operation. However, some of them do respond to the occurrences and to the environmental reality by changing their usual actions, thus configuring a dynamic component in their operation.

Let us now specify the concept of *behavior* in our context, discussing first how event-driven models and their simulators can represent and handle dynamic aspects of such processes. "Behavior" is usually understood as the set of natural or artificial reactions, exhibited by entities, to the interactions with their environment. Considering that such an environment, even when strictly inanimate, is usually very intricate, and that the structure and actions of living beings are comparatively far more complex, one concludes that behavior results in something extremely hard to define, understand, model and simulate.

From our perspective, behavior may be considered, for simulation processes, as the computational response of the model's dynamic reaction to its input. In our context, the stimulating input includes parameters detailing the inanimate environment and the stream of events the simulator receives and is expected to handle. Of course, by constructing a simulation system, one expects it will exhibit some behavior approaching that of the phenomenon being simulated, taking into account the desired functionality of the simulation and the requirements on detail, performance and accuracy intended by the user.

Most formal automata-like devices are based on some state-transition model, such as finite-state automata, pushdown automata, Turing Machines, storage-transfer-based computer models such as the von Neumann architectures, sequential execution imperative programming models, functional programming models, state charts and other hierarchically-structured devices, Petri Nets and many other devices with state spaces that are defined combinatorially. There are lots of possible combinations of devices in these categories, exhibiting behaviors that can be well associated to the handling of standard interrupts in a typical sequential electronic computer system.

For these kinds of devices, behavior is a concept that matches perfectly to externally-stimulated reactive systems described by some discrete event-driven model, which reacts to a stream of externally received stimuli by activating a corresponding sequence of specific associated interrupt handlers. Similarly, event-driven simulators map their input stream into a corresponding output sequence of reactions, which can, in turn, be interpreted as the particular instance of the associated device's behavior, in the particular case represented by the current environment's situation and the currently received stimulating stream of input events.

So, for use in the present study and its actual applications, granted the hypotheses stated before, we can adopt as a good definition of behavior the simple and practical interpretation suggested earlier: we can abstract the behavior of our simulator as being simply its output language. In other words, the concept of behavior, applied to our simulation system, is the universe of output streams it is able to produce starting from its own current configuration, the configuration of its environment, and its stimulating input event stream.

More rigorously, when an input event stream $\mathbf{s} = \mathbf{s_1\ s_2 \ldots s_n}$, $\mathbf{s_i} \in \Sigma$, $\mathbf{i=1, \ldots , n}$, is received and processed by a reactive device **D**, or simulated through some event-driven model **M**, we say that **s** is mapped onto a corresponding output stream **a**, $\mathbf{a} = \mathbf{a_1\ a_2 \ldots a_n}$, of actions that are taken in response to the input received, such that $a_i = \mathbf{M(s_i)}$, $\mathbf{i=1, \ldots , n}$. So, behavior may be interpreted as a kind of language transduction, with reactive devices **D** represented by an event-driven model **M** operating as language transducers, whose behavior consists of mapping their input language $\mathcal{L}$**(s)** onto an output language $\mathcal{L}$**(a)**, or $\mathbf{M} : \mathcal{L}(s) \rightarrow \mathcal{L}(a)$, where $\mathcal{L}$**(s)** and $\mathcal{L}$**(a)** are the sets of all possible input and the corresponding output streams, respectively.

In operation, an input language feeds the device, which reacts by executing some associated handling action. Therefore, by using the device for processing the input language, streams of stimuli are converted into sequences of output actions to be executed by the device's model in response to the input received. In other words, a device **D** maps its input language $\mathcal{L}$**(s)** onto the output language $\mathcal{L}$**(a)**. Both languages are interpreted as the collections of all possible streams that the device is able to receive or deliver, respectively.

2.1.4 Language considerations

In general, the worst case input language is $\mathbf{\Sigma^*}$, the complete (less restricted) collection of arbitrary streams of input stimuli over its alphabet $\mathbf{\Sigma}$. Whenever needed, however, adequate filters (language acceptors) may be used to impose input streams that belong to particular, less general input languages. Similarly, formal limitations may be established for the kind of input sequences allowed in the input stream, and such decision may be taken with the help of adequate filters, thus explicitly forcing the device to accept inputs belonging to arbitrarily specified particular languages.

Similarly, the transducer's output language $\mathcal{L}$**(a)**, which was previously defined to be the literal one-to-one transliteration of each input stimulus into some corresponding output action, may be in turn changed by the application of convenient output filters. It is also possible to transliterate the action codes generated as output, e.g. in order to mask undesired ones, apply language transformations to some of them, and enforce further ones in which we are particularly interested. This amounts to the execution of an interpretation $\mathcal{I}$**(a)** of the output actions, and is performed by applying Turing-powerful algorithmic procedures, thus allowing far more complex overall results. In practice, such transliteration may be taken as an interpretation of the stream of output actions, and can be done by executing routines and algorithms for handling the output stream as well.

2.2 Modeling event-driven behavior

Fig. 8.2 sketches the overall behavior of an event-driven simulator. The environment generates a stream of events that are caught by sensors and sent to an input filter, which in turn examines, chooses and separates the desired ones for being further processed by the event-driven device, by sending them to the implemented version of the event-driven model **M** of the phenomenon being simulated.

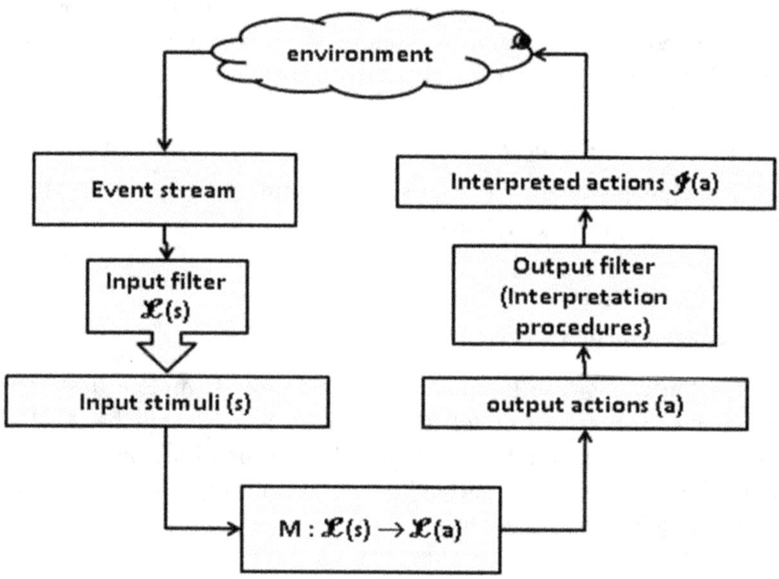

Figure 8.2: A very simplified simulation of event-driven models.

When the input stimuli reach the simulator of the implemented model of the device, they are mapped onto output actions and are further interpreted by output filters, whose results are fed back to the environment. This is of course a very simplified scheme of how such simulators work, but it does represent well almost all of the essential features of the simulation mechanism adopted.

2.2.1 Behavior in rule-modeled devices

The operation laws of any rule-driven device must be fully described by its corresponding set of rules, since "rule-driven" means exactly that the operation of such devices must be completely stated exclusively by the set of rules describing it. That is the case, for example, of the set of transition rules that describes a finite-state automaton, which specifies the actions to be taken by the device whenever some of the allowed occurrences happen at its input.

Generally, such a set of rules constitutes a very simple data structure, since it is simply the collection of all rules governing the model's operation. Supported by such a naïve model, rule-driven devices become easy to design, implement and observe, allowing them to be fully analyzed through the examination of a single data structure, namely, the set of their own operation rules.

With a model exhibiting such a feature, no more routines, no additional algorithms, no further external help is needed in order to analyze, to make hypotheses and to understand the device's movements. Therefore, the device's behavior may be inferred by tracking how the input stream drives the device's operation rules for determining the actions whose interpretation lead to the desired output procedure. This makes it very simple to obtain and analyze the behavior of a rule-driven device whose simulation is based on the implementation of an event-driven simulator supported by a rule-driven model of the device being simulated.

Under these hypotheses, behavior can be interpreted merely as the device's transfer function, meaning the mapping law from the device's input to the corresponding output. Additionally, the output represents the overall response of the system, obtained by the interpretation of the output actions. From this reasoning, we can easily conclude that the event-driven simulation of rule-driven devices is a particularly simple and natural way to obtain a suitable computational implementation for the desired simulator, since the event-driven style satisfies very well all the requirements needed for the construction of good simulators for this class of devices.

In this case, events are directly identified to the input stream elements, and their chronology may be partially or fully interpreted as some sequential law established for ordering the elements in the event list. Under those assumptions, chronology may be straightforwardly determined by simply observing the sequential order of occurrence of the events in the ordered list.

2.2.2 Event-driven simulation of rule-modeled devices

A device's *configuration* is the set of all information needed to establish uniquely all information on variables affecting somehow the simulation environment, so that by saving the current configuration of a device it is possible at any time to resume the simulation simply by restoring the previously saved configuration. Thus, given the current configuration of the device and the current input event being processed, searching the set of rules defining the device's operation allows determining some particular rule to be immediately used, and the application of such a rule allows establishing the following configuration of the device.

By iterating this process, we may go through all input events in the event list – remembering that the event list must hold all independent information needed in order to completely describe the current simulation process. Furthermore, a proper event handler associated to the chosen rule is activated, and it promotes the application of adequate output actions, according to what is specified by the currently chosen rule.

Based on a set of rules that defines the device's operation, the whole simulation process includes, therefore (see Fig. 8.3): the initialization, for the initial setting of the simulation parameters; the main simulation loop, intended at getting the next event from the input event list; and decoding, in order to select the proper handler associated to the current event and to activate the corresponding output action.

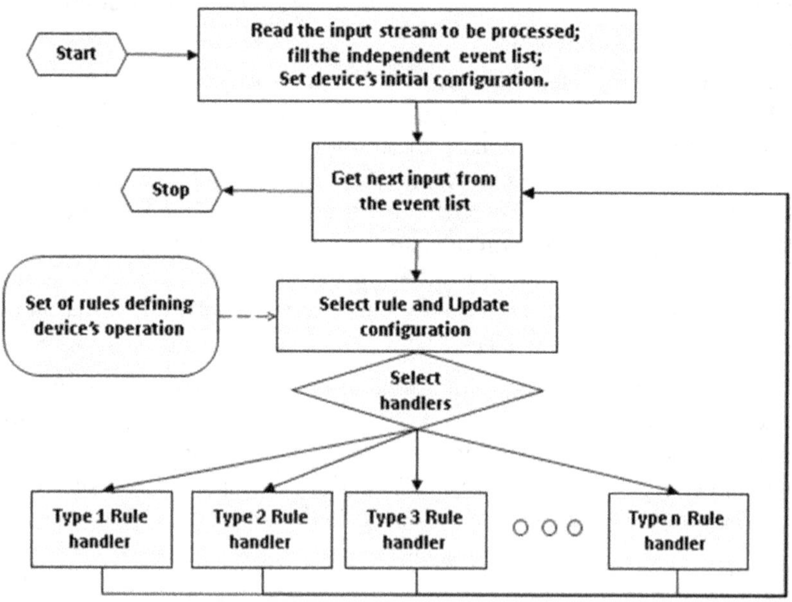

Figure 8.3: Event-driven simulator for rule-driven reactive devices.

2.2.3 Computational handling of variable-behavior models

In the sections ahead, we focus on some important conceptual aspects of the computational implementation of dynamically-changing phenomena, i.e. variable-behavior processes or devices whose operation comprises the concept of adaptivity.

The behavior of most automata-like formal devices shows a very close similarity to the handling of a succession of external interrupts in a standard computing system, therefore there is no need of extra resources for its implementation in our simulator.

3. Adaptivity issues

3.1 Adaptivity

Processes discussed so far do not change in time, so the models suggested for representing them exhibit constant operation laws, which (in rule-driven devices) imply defining them by means of unchangeable sets of invariable rules. *Adaptivity*, in the sense assumed in this publication, enables devices to change their behavior by self-modifying the set of rules defining their operation, without direct external interposition, just in response to their history and input stimuli.

3.1.1 Enabling self-modification

We will now study how adaptivity influences the modeling and simulating processes. Adaptivity allows devices to make changes to the set of rules defining them, so that any device supporting adaptivity is able to modify the set of rules that determines its own operation, therefore changing its local behavior. By changing the set of rules of a device, its response to the received input can be modified. An adaptive device is obtained by allowing the device to change its own set of rules without external help, so that the only element affecting its behavior at any instant is the device's history.

In order to enable adaptivity, rule-driven devices must perform self-modification, and that may be achieved for example by attaching to its rules special adaptive actions to be executed whenever the associated rule is applied. Adaptive actions are represented as specific procedures instructing the device about the changes needed in order to obtain the desired modification in the device's behavior. Since the device's behavior is encoded in its set of rules, then when the device has a fixed set of rules it will never be able to change its own behavior. On the other hand, giving the device the ability to edit its own set of rules turns it into a variable-behavior device.

Adaptivity is therefore easily achievable with the help of a couple of trivial editing operators, for example, one for adding new rules and another one for erasing an existing rule. So, by giving the device access to this pair of editing operators, it will gain the ability to change at will its own behavior.

3.1.2 Structure of adaptive rules

We mentioned before that adaptive rule-driven devices are specified by a set of rules which can include the so-called adaptive actions, whose application allows updating the set of adaptive rules, consequently changing the device's behavior. An adaptive rule may take a configuration as shown in Fig 8.4.

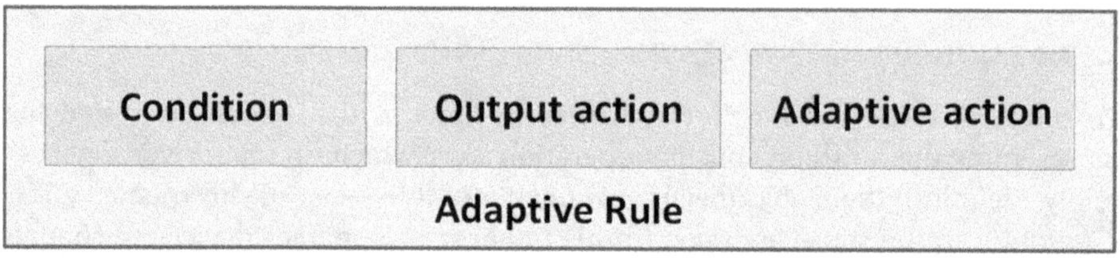

Figure 8.4: Structure of an adaptive rule.

There are three components in the conditional clause that represents an adaptive rule. (a) The condition field specifies the circumstances under which the rule can be applied. In non-deterministic cases, there may be more than one rule selectable for application. In error situations, there will be no such rules at all. (b) The second field specifies the output action to be performed when the rule is executed. (c) Finally, if a non-empty adaptive action is indicated in the third field, the specified adaptive action will edit the set of adaptive rules in order to change the actual device's behavior.

In the currently used formulation for adaptive devices, up to a pair of adaptive actions associated to each rule is allowed, one intended to be executed before applying the rule, and the other after the completion of the output action.

3.1.3 Structure of adaptive rules

The dynamics of the operation of adaptive rules is sketched in Fig. 8.5.

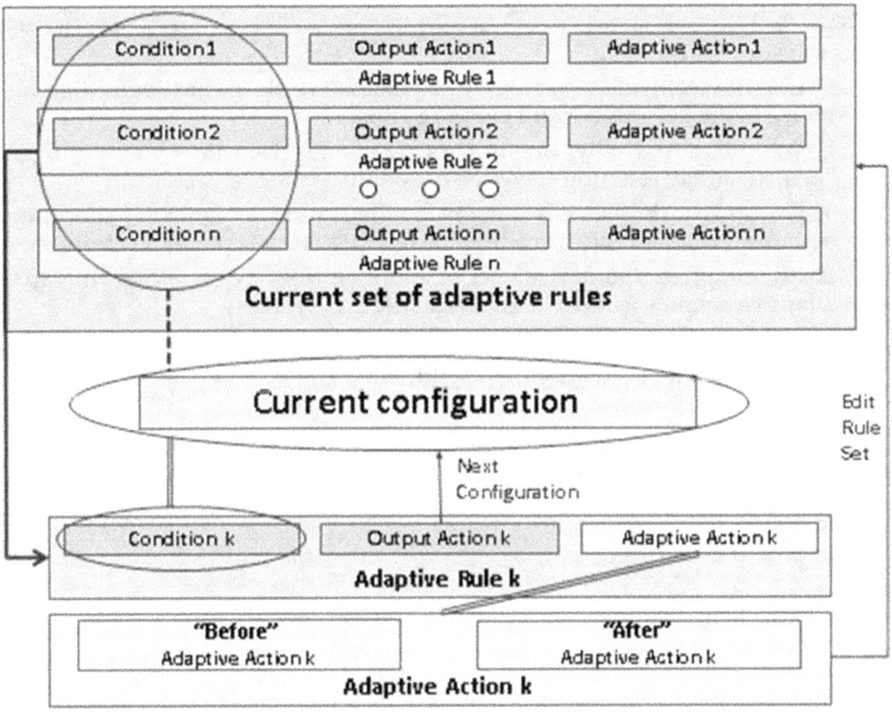

Figure 8.5: Operation of adaptive rules.

At the top of the sketch, the current set of adaptive rules is shown, each of such rules following the form presented previously. The device's current configuration assembles most variable information about its current status (excluding the set of adaptive rules). This configuration represents a kind of signature of the whole status of the adaptive device at the time it was examined, and concentrates all information needed for enabling the device to perform the next step in its operation. The information in current configuration is searched in the condition field and some rule matching the current configuration is eventually selected from the set of adaptive rules. By applying the chosen rule, the current configuration is updated to some new one, as specified by the rule. On the other hand, by executing the adaptive action(s) specified in the adaptive rule, if present, the current set of adaptive rules is updated as well, ending the cycle and enabling the device for the application of the next rule in its operation.

The logic of operation of a rule-driven simulator may be summarized in the pseudocode shown in Table 8.1.

For each simulation step:
– **Search the set of adaptive rules for a particular rule k whose condition field matches the device's current configuration.**
– **First, update (partially) the device's set of adaptive rules by executing the "before" adaptive action specified in adaptive rule k, if present.**
– **Then, update the device's current configuration by applying the output action specified in rule k, leading the device to its next configuration.**
– **Lastly, complete the device's set of adaptive rules by executing the "after" adaptive action specified in adaptive rule k, if present.**

Table 8.1: Pseudocode for the logic of a rule-driven simulator.

With such operation, adaptive rules show high compatibility with usual rule-driven devices, and the simplicity of this approach allows easy implementation of models based on it.

3.2 Working with adaptivity

When variable functionality is desired for some computational device, it may be accomplished in several ways. A few of the choices for implementing adaptivity in rule-driven models are briefly discussed below.

We will start with the non-adaptive case. Such devices operate without changing their behavior; consequently, their running laws remain invariable throughout their whole operation.

3.2.1 Invariable-operation devices

Most devices that are often found in practice do not need variations in behavior, so they can operate steadily, with no changes, remaining structurally and functionally unmodified throughout their usage. In this case, the implementation of such devices can be performed easily as rule-driven devices with invariable operation laws, which can be stated in terms of a constant set of unchangeable (read-only) rules.

3.2.2 Variable-operation devices

Certain devices, however, allow changes to occur in their operation laws while running, requiring either a variable set of read-only rules or a set of modifiable rules for their adequate characterization. Consequently, for implementation purposes, variable-operation devices may explore in several ways the possibility of changing their set of defining rules.

Some devices allow presetting their behavior by simply choosing among a number of pre-established unchangeable sets of rules, others may have their sets replaced on-the-fly, and yet others may even have the ability to modify their own set of rules.

3.2.3 Static selection of an adequate rule set

The simplest way to achieve variable operation in a computation is by statically selecting at once its set of rules, e.g. before the device is put into operation, at the time of the device's initialization, by choosing one among several predefined rule sets, as sketched in the diagram of Fig. 8.6.

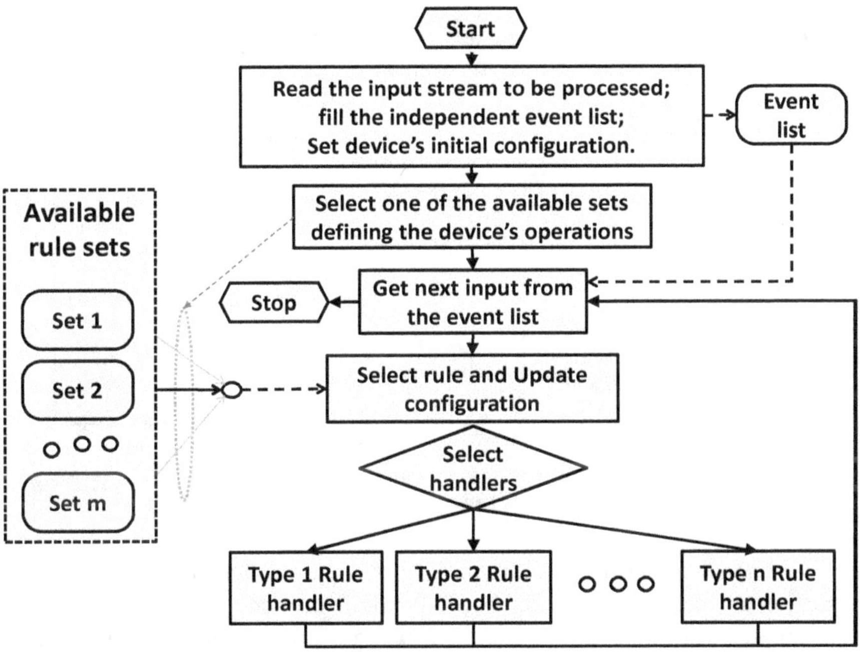

Figure 8.6: Simulation of statically function-selectable devices.

The leftmost big rectangle in the sketch shows the collection of available rule sets for the device, among which the simulator can choose one to be permanently used in the subsequent simulation. This situation occurs with function-selectable devices through the choice, at the time of device's initialization, of one amongst several predefined sets of rules. Afterwards, the chosen set of operation laws is kept unchanged until the completion of device's operation.

3.2.4 Dynamic selection of the rule set

As a next step, we can elaborate slightly by allowing the dynamical selection of a suitable rule set, to be followed by the simulation process, e.g. by allowing the program to select on-the-fly one of the device's possible functionalities. In event-driven implementations, that is easily accomplished by externally firing asynchronous events in order to specify (dynamically) the set of rules defining the desired behavior of the device at each instant of its operation.

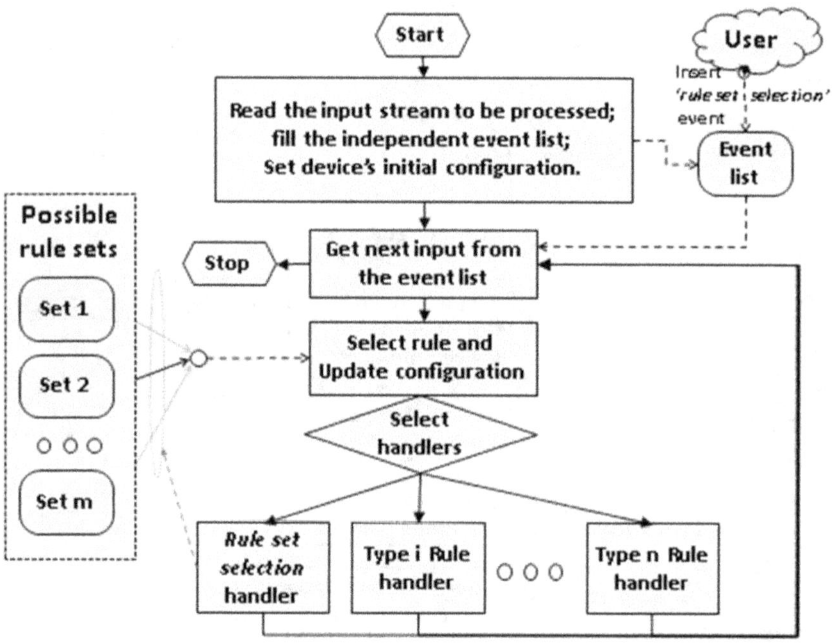

Figure 8.7: Simulation of dynamically function-selectable devices.

Such a choice brings a considerable improvement in flexibility to the usage of the device, since several functional modifications may occur along a single run of the simulator, although by the time of each change the whole functionality of the device is fully switched at once. The diagram in Fig. 8.7 shows the logic of dynamically selected switching among a predefined set of rules.

3.2.5 Self-modifiable (adaptive) devices

As may be easily observed, devices with dynamically variable operation laws (also called "adaptive devices") can explore the powerful ability of self-modification (i.e. changing their own operation rules on-the-fly). The use of adaptive devices is among the most sophisticated choices for implementing a self-modification operation, since it even allows the device to decide autonomously, without help, when and what to change in its own current set of operation rules in order to fit the needs of the application at each instant.

Self-modification ability can be achieved in a particularly simple way by only implementing a very small number of primitive editing procedures, since self-modifications on the device's operation laws (i.e. amendments in its behavior) may be always reduced to the application of editing operators to the device's set of rules. This may be done, for example, through an appropriate combination of the elements of a very small set of operators, such as {replace}, or {insert, remove}, or {find, replace}, etc. Fig. 8.8 shows an outline of the logic for an event-driven simulator of adaptive rule-driven reactive devices, which operates in accordance to the patterns discussed above.

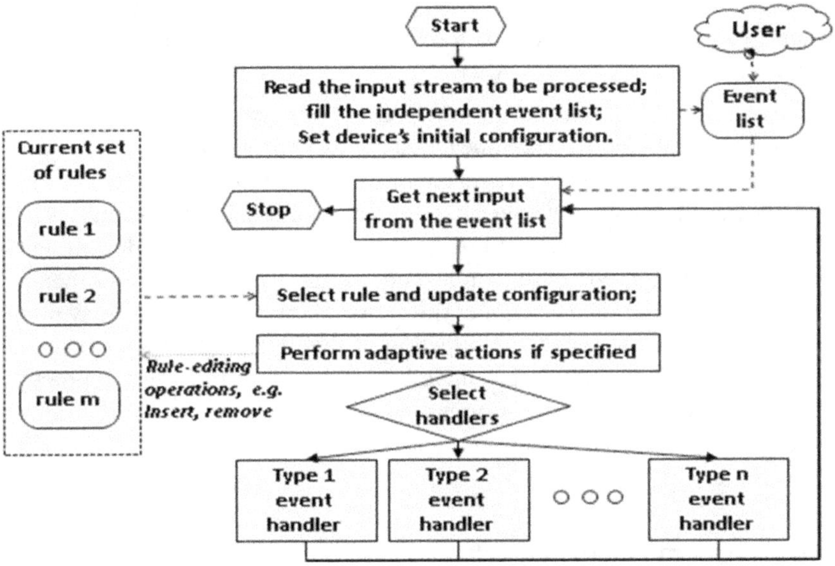

Figure 8.8: Simulation of self-modifiable (adaptive) devices.

3.3 Applications of adaptivity

One can use adaptivity in a variety of ways in programs such as a complex system simulator, and the remaining topics in this text mainly refer to a diversity of ways the concepts mentioned above can be employed, with emphasis on how to apply adaptive techniques in the construction of rule-driven discrete-event-driven simulators. It is important to note that this proposal leads to an adaptive software, which, as any software system, needs discipline for its conception, design and implementation, particularly in regard to the self-modification issues. We can find in the literature some surveys covering works on how Software Engineering deals with this kind of software systems (see Andersson et al., 2009).

3.3.1 Using adaptivity in simulators

Being a complex program, the simulator itself may be encoded as an adaptive program, by following one of the proposed patterns described above or any other equally proper techniques. In programs of this category, there are often a lot of different techniques that can be indifferently employed to achieve the needed simulation purposes. The adequate choice among a multitude of combinations of available simulation algorithms and policies constitutes itself an intricate decision problem, whose complexity may itself recommend the use of adaptive techniques.

Usually, complex programs and their users require complicated graphical and/or distributed, multi-user interfaces, for which significant features to be considered are the visualization mechanism used, zoom features, granularity, resolution, need of rendering image details, etc. Additionally, in systems with such a complexity it is important for the user to be able to select the desired level of simulation in order to mask undesired details, therefore simplifying the computation required. The control of the need of simulation details may often be done automatically, as well as the selection of significant, although considerably smaller, subsets of the simulated universe, in order to limit the attention (and consequently, computational resources) to predominantly handle the currently most relevant part of the simulation.

3.3.2 Extra applications of adaptivity

Numerous applications can benefit from the use of adaptivity, since in many circumstances it may be helpful for improving the efficiency, accuracy, and computational performance of co-operating methods. Among general applications of adaptivity, the most relevant are those related to artificial intelligence, such as machine learning, information mining, knowledge acquisition and organization, decision making (automatic or assisted), process optimization, and many others.

Efficiency can be improved especially in cases related to combinatorial problems, in which adaptivity can help to reduce tree-spanning-originated exhaustive computations by trimming the tree in favorable conditions, thus avoiding permanent exhaustively searching all possible non-visited combinations. Accuracy may be enhanced mainly by using adaptivity for determining whether the current situation corresponds or not to reaching the limits of benefits that some method or algorithm is able to deliver. In such a case, adaptivity can also help to implement the automatic selection of another technique, method or algorithm (or another combination of them), that would adequately replace the current one in the search for more accurate results. Computational performance may be a very complex combination of (usually numerical) quality indexes associated to attributes of the process being evaluated. Such indexes are often figures evaluated by applying suitable qualitative or quantitative metrics to the measured phenomenon.

There are many other relevant aspects to be considered concerning the impacts of adaptivity in technology, but they will not be addressed here, since they fall outside the scope of this qualitative and introductory study.

3.3.3 Adaptivity and simulation

Even under the limitations present in the (although somewhat advanced) modern technology, the combination of adaptivity with other techniques, such as simulation (and/or with many well-known classical techniques as well), turns out to be helpful in the elaboration of novel and more effective solutions for complex problems. As said before, simulation has been the illustrative application chosen here, for its immediate interest for this text, even though any other similar area of interest could equally suit such purposes.

The combination of adaptivity with simulation is explored here as a possible and attractive way to enrich simulators with some properties that adaptivity is able to add to methods or techniques which it is attached to. Such combination can enable simulators to handle, in a natural form, models whose behavior can somehow vary along their operation. Therefore, the blend of adaptivity and simulation becomes here an important question to be considered, since adaptive features are very important in the style of simulation we are examining.

It is appropriate to remember again that adaptivity is not to be considered a dogma that should be unconditionally and inflexibly followed. On the contrary, it constitutes a form of reasoning that allows either replacing others that do not fit the needs of the current application, or complementing them by trying to improve those attributes they cannot satisfy alone, in order to fine-tune them in compliance with the requirements of the current application.

3.3.4 Adaptivity can upgrade standard techniques

Standard divide-and-conquer strategies usually help one to separately investigate, understand and explore the multitude of aspects considered of interest in a problem whose solution we are concerned with. This technique is also known as “successive refinement”, and it allows to recursively decompose, incrementally, the original problem into easier ones, until all sub-problems created in this refinement process result in trivial ones, i.e. problems whose solutions are already known. Every time this technique is applied to some initial problem, that problem is decomposed into a set of sub-problems along with some additional logic, such that the resulting ensemble is equivalent to the initial problem.

Adaptivity is particularly well-suited for working together with other techniques that use divide-and-conquer strategies, since processes that combine both are not required to develop entirely the refinement, if already refined parts are ready to be executed. Programs that use this approach are also known as “lazy-execution programs”, since the completion of their programming process seldom occurs before they are actually put into operation, so undeveloped parts are expanded strictly upon request, when its execution is required, remaining unexpanded otherwise.

The control of the expansion process in lazy-execution systems is well handled with adaptivity, since they can be designed for being lazy-execution systems. Evidently, there are many other techniques beside lazy-execution and successive refinement that fit all the requirements to work well together with adaptivity, but we have no space here to extend this text for suitably discussing them.

Another comment may be useful to complete the idea in the last paragraph of the preceding section: excluding uncommon situations, adaptivity is better suited to work in cooperation with other techniques than to being employed in place of them. This makes it wiser not to consider putting adaptivity in competition against other techniques, but using it instead for complementing them, so as to fulfill each other's weaknesses whenever possible in order to get the best features from both worlds.

Finally, it is very important to note that adaptivity is not intended for use as some kind of “magic” for solving all problems, since it has been demonstrated that adaptive devices are Turing-powerful, therefore they can in principle solve any Turing-computable problem, although not being expected to solve any problem that a Turing Machine cannot handle. Although adaptivity does not add any computing power to classical Turing Machines and all their equivalents, adaptive devices are themselves a good option for handling computations, and one may consider using them because of possible advantages they exhibit in certain situations over other available choices.

We have already commented on some interesting uses of adaptivity in conjunction with other methods and techniques, and this seems to be a promising way to make incursions into unexplored and hard to access fields of knowledge, such as one finds in many branches of cognitive science, cosmology, bio-sciences, physics, etc.

4. Issues on behavior

4.1 Modeling complex behavior

Our purpose is to discuss some topics on the characterization of behavior in the computational representation of complex phenomena, so as to collect specifications for the implementation of a computer simulator for some phenomena being considered. See Langton (1998) for more considerations on this issue.

In order to simulate a complex phenomenon in a digital computer, it is essential to study the properties of the phenomenon, in order to acquire enough knowledge for conceiving and building a formal model, and then, to convert the model into a logical device, able to be encoded for execution as a computer program.

In this section, we investigate some requirements for modeling behavior, and try to establish a few of them for such models. Then we study characteristics for populations and their constituents, the structure of a model for an individual of that population, and conclude by examining some behavioral properties of the individuals in the population, and by proposing an action line for modeling and implementing such models as an adaptive device for being used together with the simulation scheme studied before.

4.1.1 Considerations on modeling behavior

In order to evaluate in a more concrete way the use of adaptivity in simulation, so as to naturally include behavioral elements, it is convenient to illustrate with a case study concerning the simulation of any natural or artificial being with any kind of complex cognitive ability. We chose a case study involving the simulation of a living being, for which we make a first move toward specifying qualitative requirements.

In this section, we list some desirable elements for modeling/simulating the actions of living beings, related both to their individual and gregarious behavior. For individual beings, there are lots of aspects to be considered, many of which are very complex: inheritance, birth, growth, nutrition, life, health, reproduction, aging and death; dangers, security, defense; ideal physical conditions for life, forms of survival; reaction to external stimuli; proactivity in response to the specific needs.

For collective gregarious beings, in addition to the information on each individual, it is necessary to consider: family relations, the logic of their community organization, the way individuals seek their social prominence; how individuals achieve their roles in a community, how they can establish relations of dependence and collaboration; how individuals in a community relate with others in order to establish relations of awareness and competitiveness; how they exert actions for improving their social culture, by acquiring knowledge, by growing their intellectual, philosophical, moral and ethical values, and many others.

Our scope in this text is far from taking into consideration so many complex subjects, so we limit our discussion just to some exploratory shallow considerations on a small number of the simplest of them.

4.1.2 Toward modeling behavior components

Modeling behavior at least means modeling the enormous variety of elements that compose together its countless intricacies. The first step consists of collecting a significant set of attributes that is enough to distinguish the organism, and therefore must be included in any model representing the simulated creature in the computer (in our example, a living being).

Besides their own individual behavior, gregarious living beings exhibit a partial behavior related to their role as a member of their community. Part of this extra behavior refers to the influence of the community on the individual behavior, and another part, to the individual's role in building the community's joint behavior.

The most important elements the model must include are those related to features of our living being, considering: physical attributes, such as dimensions, age, position, speed, weight, temperature, etc.; biological attributes, such as life cycle, health conditions, food restrictions, immunological conditions, pharmacological needs, hyper-sensibilities, sensorial restrictions, etc.; psychic attributes, such as: intelligence level, skills for specific kinds of activities, preferences for specific activities, creativity, ease for combining facts, knowledge acquired previously, ease for communicating, memorizing, understanding and learning; communitarian attributes, related to the qualities that the individual assumes (even unconsciously) while belonging to some organized society.

4.1.4 Individual physical characteristics

Among cognitive characteristics, it is desirable that the following be considered for individuals: understanding of the environment and recognition of its components and their relationships; identification of individuals and their relationships; communication among individuals; understanding messages from other individuals; memorizing information; organizing data; interpreting data; learning; modeling knowledge; extracting knowledge from data; organizing knowledge; acquiring and memorizing time information; managing laws for information retention and forgetting; using conditioning, bonus and penalties in learning techniques; requesting and delivering knowledge to others; recognizing and solving problems: curiosity, identifying doubts, formulating questions, searching for answers; satisfaction with learning, socialization of gathered knowledge, collective knowledge training; awareness, ethics, reasoning, emotion, dreams, beauty, art; pattern recognition, classification, decision making, selection of criteria, policies.

Among collective characteristics of gregarious beings, the following are interesting to be present in the model: knowledge of the organization of the community, consciousness on how to proceed in order to achieve a satisfactory social position; consciousness on how to seek a suitable social prominence, consciousness on the individual's roles in the community, information on each individual's dependence on the others; consciousness on how to adequately collaborate with pairs; consciousness on the need to compete with others in order to achieve some special goals in the community; collective behavior; collective consciousness; ethical values; intellectual, philosophical and moral values; knowledge, culture.

4.1.5 Toward modeling individuals

In the sequence, we illustrate some possibilities of implementation by briefly discussing how adaptivity can be employed for handling a short subset of the large list of complex phenomena mentioned before, in order to gather information for helping to start the elaboration of a mathematical model adequate for the construction of simulators of complex phenomena such as an individual living being.

Obviously, any attempt to simulate complex systems and living beings with a minimum of accuracy requires the implementation of a lot of methods and operators representing components which may help constructing hard-to-implement operations. Nevertheless, all of them are always mere approximations of (the usually far more complex) reality, since very often natural phenomena are extremely hard to understand and simulate.

4.1.5.1 Reaction to the environment

One of the most important features of living beings, which must be implemented in any effort to imitate their natural behavior, is the ability to react to stimuli received from their environment. Such reaction to the environment is a primitive part of the behavior of living beings, since they must, at least to survive, feel their environment, learn how to collect detected stimuli and received sensations, and classify them, in order to correctly respond in a specific way to each possible sensation received.

For this reason, it is fruitful to use, in the modeling and simulation process for these phenomena, techniques directed by events in computer simulations of reactive models, since with this choice the formal model and the simulation of this aspect of the behavior of the elements simulated become very robust.

A concrete proposal for a natural accomplishment of reactions to the environment in a simulation background has been previously discussed in detail, in several diagrams proposing implementations for event-driven simulators for reactive phenomena represented as rule-driven models (such as Fig. 8.3). Hence, reactive models can properly represent reactive systems, and simulating such models clearly fits our needs for a good approach to implement these features.

4.1.5.2 Instincts

Living beings are born ready to perform certain complex tasks, even without having ever learned them, and this ability provides them with the so-called "instinctive behavior".

Instincts may be reactive or proactive. *Reactive* instincts provide the living being with responses to environmental conditions it is able to sense, while *proactive* ones allow it to anticipate its responses to the environment. While reactive instincts provide causal reactions, proactive ones do not need any stimuli to spontaneously produce random initiatives generated by the simulated living being. Reactive instincts may be considered in the simulation process by means of the inclusion of a predefined subset of the initial rules. This portion of the behavior can be easily inserted into the simulated environment by pre-including (before starting the simulation of the device) an appropriate set of initial rules that perform such tasks. Randomness, needed for the simulation of proactive instincts, can be simulated with the help of randomly choosing both the timing and the activity to be performed.

Since instincts (at least some of them) are inherent to individual animals throughout their lifetime, a natural way to represent instincts in the simulation environment is to include in the device's initial set of rules a predefined subset of rules that manage particular events from the environment that stimulate specifically the corresponding instinctive reactions. By applying such rules, the associated adaptive actions carry out simulated reactions that may even perform changes in the adaptive device's set of rules.

Another concrete suggestion for simulating animal's instincts is to model them by means of an adequate subset of the rules of the adaptive device, since the device's behavior at the instant of birth of the simulated animal must be fully described by its model's initial set of adaptive rules. A properly timed activation of a collection of adaptive actions can take into account variations that occur in its instinctive reactions throughout the animal's life, according to its age.

4.1.5.3 Intuition

Different from usual history based events, an intuitive component of behavior usually has no obvious relation with its causes, and can be modeled by promoting spontaneous response actions to intuitive stimuli. In order to model intuitive behavior in our simulation, we can impose the schedule of special high-priority impulsive intuition events, to be handled as soon as possible. The handling of some regular events can potentially schedule such random impulsive intuition events at arbitrary yet realistic instants. The handling of impulsive intuition events causes the execution of actions which can be interpreted as a reaction to the corresponding intuition.

4.1.5.4 Proactivity

Proactivity is a constituent of behavior that enables the elements being simulated to spontaneously perform non-stimulated activities. Viewed as a computational action, proactivity may be considered as a phenomenon of the same class as the aforementioned intuitive, instinctive and reactive responses.

In our event-driven simulation process, implementing proactivity allows us to establish occasions in the simulation process at which additional stimuli can be freely inserted into the input list. Subsequently, such added stimuli are supported exactly the same way as any others are, and may be handled by the simulator through the generation of usual events associated with the desired reactive responses.

In order to model proactive behavior in our simulation, it is sufficient to allow scheduling proactive events at appropriately chosen instants of the simulation. The handling of any event can potentially schedule (either randomly or at simulation instants selected according to some arbitrarily chosen criteria) proactive events at arbitrary instants.

4.1.5.5 Dreams, nightmares, hallucinations...

Living beings usually experience periods in which they temporarily have the sensation of belonging to some parallel world with specific features, often different from the real ones (see Tchitchigin et al., 2016). Such occurrences are usually sensed as dreams, and take place while sleeping. Others (nightmares and hallucinations) are frequently associated to physical or psychological disorders, and can cause particularly strong and disturbing sensations.

From the simulation viewpoint, they may be classified in each category according to the type and the strength of the sensations, and essentially do not differ in most aspects, which allows us to implement them parametrically, exactly the same way. Dreams, nightmares and hallucinations can be simulated by means of a simple recursive call to the simulator for a chosen interval of time, provided that it uses an adequate set of rules that matches the needs of the particular parallel world for that simulator instance. At the time the simulator is activated, the duration of the starting dream can be established by scheduling an "end-of-dream" event to occur at some desired instant in the future. The dream is over when the simulator handles an "end-of-dream" event, which provides a return to the previous instance of the simulation procedure, which had been activated with the set of rules associated to the model of the actual world.

4.1.5.6 Memorization and learning

Programs extensively assume some representation for information, acquire, encode and store relevant information extracted from facts that already occurred, in processes of memorization. As we have mentioned before, in our simulator, the representation chosen for all information must be completely and exclusively encoded in the set of adaptive rules. Further analysis may be performed in order to organize the collected information, eliminate redundancy and infer non-explicit facts from already learned information. In accordance with the adopted representation, all this activity is also concentrated in the adaptive set of rules.

In our simulator, implementing memorization and learning operations manipulates exclusively the set of simulation rules, and no additional information is used. Therefore, no extra computational costs are imposed on the simulation process by such operations. Being implemented solely through the adjustment of the set of rules, memorization and learning are made exclusively through the application of editing operations to the set of rules (e.g. by including and removing rules only). Rules expressing novel information are merged into the current set of rules that describe the adaptive device being simulated. A classical case study illustrating the process of collecting and representing information with the aid of adaptive devices is the word collector (Neto, 2002).

4.1.5.7 Decision making

Throughout their existence, living beings must make lots of decisions, some of which vital, particularly when they need to make choices based on their current conditions and of their background. In our simulator, decision processes can be implemented by applying any conventional technique, for example by employing (possibly adaptive) decision tables or decision trees. Models based on adaptive devices can make technical decisions in a very natural way with the aid of such devices (Neto, 2002).

4.1.5.8 Pattern recognition and imitation

One of the most important tasks that allow simulated phenomena to make decisions is the recognition of patterns. For living beings this ability may signify their survival. One important recognition ability among living beings refers to the perception of the patterns followed by other beings in their community, since imitation allows acquiring and reproducing (perhaps in a variety of contexts) already experienced behavior patterns without needing to repeat the whole (usually lengthy and very complex) learning process.

In our simulation process, (behavior) patterns may be acquired for example by using machine learning techniques, and their context may be identified by means of pattern recognition procedures. Once recognized and acquired, a pattern may be represented and collected in a way similar to that cited above used for collecting words (Neto, 2002).

The same device may then be employed to implement imitation in the simulator: it is a rather easy task to transform the device representing the collection of patterns already learned into a generator of instances of those patterns (such use is very frequent in sentence generators used in computer-assisted courses, or in automatic test generators, used in software engineering program-development environments).

In some specific applications, such as OCR (optical character recognition), image diagnostics, classification of geometric figures, and many others, patterns are not rigid, but allow significant variations, and in such cases, it is convenient to incorporate distortion operators in order to enable such variations in the shapes. In particular, imitation in different contexts requires some decision-making for determining the variations tolerated in each situation, and needs a further effort to adjust (by allowing the application of adequate changes to the rules of the device) the available collected patterns for adequately fitting their actual context.

4.1.5.9 Individuality and collectivity

For a complete, thorough simulation of gregarious beings, in addition to the particular behavior of individuals, both their behavior as members of their community and the community's behavior as a single body must be considered.

The particular conduct of individuals is of course the fundamental behavior to be simulated, and usually takes into account the presence and the interaction of other individuals and other components of the environment. The simulation of an individual's behavior is therefore based on the individual's reactive response to all other elements in the environment, all of them considered (at least in a first approach) as being independent sources of external stimuli.

On the other hand, the behavior of individuals as members of their community must consider in addition some important changes in the behavior when individuals are not alone but are accompanied by others instead. In this case, all of them receive influences from each other, producing some kind of collective behavior, which never occurs with solitary individuals, and in practice is usually very different of the individual's behavior. Another curious mutual effect of individuals in swarm-like communities on each other's behavior refers to the resulting (in general, unintentional at the individual's level) composite behavior of the community as a whole.

For all such cases, we can apply the concept of hierarchical organization of the models used in the simulation, especially those employing adaptive devices with hierarchical structure, which fit successfully the needs of this class of simulation by allowing simultaneously modeling various behaviors, one for each simulated level of social relation.

4.1.5.10 Rewards and penalties

Training, conditioning and learning are very important applications of simulation, but they are also handy for use in the simulation process itself, in particular when implementing models of natural behaviors. Rewards and penalties are simple and successful techniques that nature uses extensively for building and shaping the cognitive behavior of living beings, and the simulation of such behaviors can be particularly well achieved with the aid of classical techniques, such as reinforcement learning.

In nature, living beings are separately rewarded or punished for their acts, leading individuals to a permanent search for defense and gratification, and a continuous rejection of situations in which they can be someway penalized. Nature usually rewards desired behavior by exciting the brain with some chemical or electrical stimulating reaction that produces some sensation of well-being or pleasure. The reverse is also true, since undesired behaviors are habitually punished somehow, e.g. by inducing disgusting sensations of pain or discomfort.

By repeating consistently such rewards and penalties in response to practical experiences, the final effect tends to adjust the correct conditioning of the individual to prefer the practice of desired behaviors and to avoid undesired ones, since the individual instinctively tends to select conducts with high levels of reward, with preferably no punishment. In order to simulate such a behavior in a computer, it is possible for example to imitate the natural effect of pheromones, endorphins and others natural active substances on the behavior of living beings.

In our simulation, reinforcement learning, as well as any other classical Artificial Intelligence techniques, can be freely applied, alone or in conjunction with adaptive methods, for modeling behavioral aspects of simulated beings, in particular those related to learning phenomena. In our approach, the effect on simulated beings of active stimulating elements present in the environment may be implemented just in the same way simulated reactions are produced in the simulation process, as a reaction to received events.

4.1.5.11 Birth, death and reproduction

All living beings have finite duration in nature. They are born, they live and they die. In order not to extinguish quickly, communities of any species must at least replenish with new individuals, while others exhaust their lifetime and disappear. Living beings must therefore reproduce, otherwise their species would extinguish. In nature living beings exhibit two basic types of reproduction: sexual and asexual.

Asexual reproduction is the simplest one, but it is bound to the creation of new instances of an existent living being. Cloning is one such form of reproduction, very common in plants but rare in animals. In primitive unicellular beings, for example, this kind of reproduction is performed by splitting the original individual, and by making a copy of its genetic material. This phenomenon does occur somewhat frequently even among mammals in their earliest stages of pregnancy: identical twins result from cellular division, just the same way as with unicellular asexual reproduction. In plants, asexual reproduction is very frequent, and takes place by regeneration, producing a whole body even from small parts of an existing one. In certain species, this is even the only mode of reproduction for some varieties whose evolution led to a form that has no more sexual reproduction capacity. Asexual reproduction in adult animals is quite rare, but it does occur in certain primitive species, such as planarians, that have the capacity of regenerating removed parts of their body. In other uncommon species, this regeneration capability does not allow them to reproduce, but is restricted to the regeneration of certain parts of the body only.

Sexual reproduction is much more complex, since it involves a pair of opposite-sex partners, and provides genetic transformations, improvements, adaptation and evolution through a highly intricate combination mechanism applied to their genetic material. Artificially, computation has in the last years put much effort in this area, motivated by genetic engineering and research on reproduction, diseases and evolution.

Nowadays there are plenty of bio-inspired methods and techniques, all of which extremely complex, that may be employed in many applications involving situations in which nature turns out to be the ultimate source of effective inspiration. The simplest of the advances in the direction of imitating natural behavior, although in a fairly naïve form, is the genetic algorithm technique, one of the earliest bio-inspired techniques adopted in Artificial Intelligence.

4.1.5.12 Handling too complex behaviors

In the current state of technology, a yet ambitious challenging overall goal remains steady in focus, despite its complexity and low feasibility: replicating the behavior of very complex processes, such as ecosystems, cosmic phenomena, subatomic processes, biological bodies, and especially processes related to brains, such as reasoning, cognition, intelligence and behavior. However, even though the whole intent is too hard to accomplish in each circumstance, as a pragmatic tactics, we may restrict our target to some feasible (and usually very small) subset of the original objective, in order to approach, progressively and incrementally, more and more complex sub-problems, following a bottom-up divide-and-conquer approach.

Although such a strategy obviously does not directly lead to the whole desired result, it allows instead one to state a progressive sequence of feasible increasingly complex intermediate goals towards the main target, thus providing an alternative way for economically obtaining acceptable partial results, whilst the ideal desired goal remains yet unaccomplished. Obviously, the above analysis refers to the use of the currently available limited technology for the solution of huge, complex, quite unapproachable problems. Our situation may be compared to that of Artificial Intelligence when trying to give solutions to high complexity problems, and the most realistic approach here is also to relax some requirements by accepting suitable approximations of the ideal solution for the original problem. More discussion on this theme, with focus on artificial life, can be found in Langton (1998) and Bedau (2003).

Therefore, one feasible adequate scenario for the development of a simulator for our application is achievable by imposing restrictions to our model, such that only a small fraction (a "window") of the whole universe of interest is considered, and correspondingly constructing the simulator out of a collection of quite independent pieces of the model, in a bottom-up fashion.

4.2 Reflection on behavior modeling

In this work, we assumed as a primary hypothesis that our simulation is based on a rule-driven model of a reactive phenomenon, implemented with the aid of adaptive techniques, and interpreted by an event-driven simulation engine. With such a hypothesis, the most important result, cited before, which is extremely convenient for our purposes, is that the whole knowledge contained in the simulated device is encoded exclusively on the set of rules describing its operation.

Consequently, all details of such device's operation may be tracked accurately by sequentially enumerating the device's operation steps, which are in turn triggered by the occurrence of its sequence of input events, stimuli, interrupts, input elements, categorized tokens, etc. In response to such an input, the device reacts by outputting a sequence of configurations that expresses the device's movements, and a sequence of changes in the set of rules commanding the operation of the (adaptive) device.

In addition to these premises, it is easy to verify that the achievement of the basic features required in our approach to implement the simulator does not demand any extra efforts, since reactions to input stimuli, the most important constituent of our implementation approach, represent exactly the main goal of the action desired for our simulator as well.

4.2.1 Evolution of instantaneous performance

Living beings exhibit reactions that vary constantly throughout their existence. In a similar fashion, computer devices can also reveal a conduct that can change according to the knowledge acquired so far by the device. Therefore, devices exhibit instantaneous performances that may evolve while they operate, e.g., information dynamically received by the device may induce successive reactions, manifested as outputs, together with adjustments in its set of operation rules, and the consequent knowledge such rules encode as well. Rule modifications can produce side-effects that usually cause additional changes in the corresponding reactions of the device, which in turn can originate an evolutionary step in the device's instantaneous performance.

Since this effect results as a direct outcome of the techniques we adopted for performing our simulation task, it may thus be considered a natural effect of our implementation; therefore, also in this case, no further specific technical effort is needed in order to achieve our goals.

4.2.2 Behavior and its qualification

In nature, living beings are usually characterized by their behavior, meaning one's own conduct, their way of working or functioning, their reaction or conduct in response to some situation or stimulus, etc. In computer models, behavior is a concept that may be associated with the dynamic conduct or reaction of some class of devices to situations or stimuli, according to certain (arbitrarily chosen) classifying criteria.

Behavior should synthesize all significant (dynamic) properties of the device, independent of particular cases or special circumstances to which the device is submitted. In our context, behavior should be viewed as an abstraction similar to that represented by equations describing the motion of solids in classical mechanics, by explicitly establishing relations between position, forces, speed, mass, acceleration, time, etc.

In order to represent behavior for some phenomenon being simulated, our model uses the set of rules of the adaptive device instead of analytical formulations. With such a choice, since there is no restriction for the choice of such encoding, users actually have a wide range of possibilities for encoding any information in the set of rules.

When the desired behavior resembles some mathematical phenomenon, the best way to encode them is of course by rearranging rules in accordance to the transformations established in the corresponding mathematical model. However, a lot of complex phenomena we wish to simulate exhibit behaviors that are too far from being expressible analytically, even mathematically, and frequently in such cases algebraic or statistical approximations do not accomplish suitable results. Unfortunately, in such cases, usual techniques hardly deliver effective results, and thus we should proceed perhaps by trying to reapply recorded information previously gathered from similar successful real cases, or by using previously recorded historical data to generate some initial value for being used as a seed in some weakly-oriented search, or even, in the absence of better alternatives, by using some kind of blind arbitrary trial-and-error attempt.

4.2.3 Problems, languages, solutions

From the theory of computation (Lewis & Papadimitriou, 1998) we know that formal languages are well suited for use as good models for computational problems, therefore their associated formal acceptors/transducers may represent very well procedures that can handle computable the problems.

Given some data sequence, representing the input data for some problem (or a corresponding algorithm), we may interpret its acceptance or rejection by an equivalent acceptor or transducer as the classification of such input sequence by the acceptor, which answers, respectively, whether the given input may be accepted as a solution for that problem or not. Thus, for any input sequence, the corresponding output (including information on acceptance/rejection) produced by some transducer for that language may be considered as a suitable formal representation for some solution of the original problem.

Of course, theory tells us that some problems are unsolvable, i.e. no solution can be achieved for them, and also that others may have plenty (possibly an infinity) of different valid solutions. So, users have the choice of programming transducers in order to search, restricted to some selective restriction: for any of the possibly existing solutions, or to find out all of them (if finite), or to list some finite subset out of an infinite or a too large set of alternative solutions.

In the following section, we present a brief study of some concepts that are important for our purposes, focusing on reactive modeling and event-driven simulation of behavior, in an adaptive context.

4.2.4 Knowledge

All information available to the device must be encoded exclusively in the set of rules that defines its operation. This set of rules is unique and is the only materialization of all knowledge available for the device to work with along the simulation.

Encoding knowledge into the set of rules may be achieved by any classical approach, the same way used by Artificial Intelligence for representing, mining, acquiring, storing, handling, removing and duplicating data, information and knowledge. In the present case, knowledge can be formally interpreted as some function that uses the set of rules as its domain, and can apply operators capable to search, insert, remove, compute properties, combine properties, etc., and extract information (encoded as properties) from stored data, and knowledge (also encoded as properties) from stored information.

Note that whenever the set of rules changes, it is likely that the knowledge it contains also changes, and conversely, in order to modify the knowledge encoded in the rule set it is necessary to modify the set of rules. Consequently, when any additional knowledge is encoded in the set of rules, one can associate it immediately to its effects on the set of rules by establishing the connection among the added information, the cause of its insertion into the set of rules, and the exact subset of rules directly affected by such modification. This allows, at least conceptually, accurately tracking any progress in the overall knowledge acquired by the device along all of its operation, together with the corresponding information about the origins of that variation, and the precise elements of the rule set associated to that memorization.

Such contribution of adaptive techniques is exceptionally rare and valuable, since it allows observing and revealing unknown cause-and-effect relations, which is a particularly precious tool in information-mining activities.

4.2.5 Knowledge learning and forgetting

A computer process is said to "learn" new knowledge when it gathers some novel information and encodes it into its current database, and that it "forgets" some knowledge when for some reason (e.g. variations in the environment; changes in the comprehension of facts; modifications in the knowledge of the world; progress in mental processes; aging of information; gap of a long enough interval of time in living beings, etc.) that knowledge becomes useless, obsolete or unimportant, and by consequence it must be removed from the database, or adequately updated.

In the approach adopted for our implementation, the concept of learning, when applied to computer processes, is understood in the following sense: knowledge is incrementally extracted by the device from the input data being received and is progressively encoded into the set of rules describing the operation of the device at that time. So, we can associate the concepts of learning and of forgetting knowledge with the variation (respectively increase and decrease) of the amount of knowledge found in its set of rules, due to the insertion or removal of encoded information, as a side effect of operating the device. Therefore, learning and forgetting are operations that can change the amount of knowledge available for use by the device, in the form of a set of rules encoded according to some established convention.

Consequently, in our case, behavior learning and forgetting correspond to modifying the current set of operation rules of the device by executing, throughout the operation of the device, adaptive procedures that perform learning or forgetting actions, which adequately insert new rules, remove existing ones and adjust others in order to fit new situations.

4.2.6 Some ideas for evaluating the behavior of adaptive devices

In our perspective, behavior refers to the conduct, manner, and attitude of a device toward a received input. In other words, the concept of behavior associated to a device denotes the whole response given to its received stimuli, including the output it produces upon the termination of handling its input. The concept of behavior should refer to an observable feature of the device, describing its overall feedback to any input received, indicating how the device's attitude toward the received stimuli determines its responses as some mix of actions and reactions.

Instantaneous behavior of a device at each instant may be understood as the result of the accommodation of the device's set of rules to its recent history, and corresponds to the current attitude of the device toward its input, i.e., the response exhibited by the device due to the action of its present set of rules at the current instant of time.

Considering our implementation approach, in which the model of the phenomenon is implemented as some dynamic device described by a rule-driven adaptive transducer, we can easily note that, at each instant in time, the behavior of the modeled device is precisely represented by the language accepted by its non-adaptive subjacent device. In the framework of the present text, the concept of total behavior of a device refers to something that typifies the conduct of the whole device, without regard to time or to any particular input being processed.

A way to abstract both time and input is to consider the total behavior of a device as the full effect of applying the device's functionality to its input language, i.e., by solely associating the device's total behavior to its overall output language. Therefore, the total behavior of a device can be understood as the complete set of patterns it can generate as output in response to all possible sequences of input stimuli. Thus, from the viewpoint of formal languages, the total behavior of a language transducer may be interpreted as the transducer's output language, i.e. the language the transducer produces as a result of processing (translating) its input language. So, as one would naturally expect, we conclude that with this definition, the behavior of a transducer (a translator) is exactly the translation of its input language into its output language.

Another important concept may be defined at this point of our reflection: the concept of *behavioral evolution* of the device, which refers to the law of variation of the instantaneous behavior of the device throughout each run. In other words, for each (non-persistent) run of the device, the adaptive device modeling starts with its unique initial subjacent device, whose instantaneous behavior will further change at each adaptive step of operation, as a consequence of the dynamic application of adaptive actions. Therefore, after the occurrence of each adaptive operation step, a correspondent evolution will occur in the set of rules describing the device, so at each variation in the set of rules (i.e. at each execution of some adaptive function), there will occur a corresponding variation in the behavior of the device.

The law of variation of the device determines the device's behavioral evolution, which may be easily measured step by step simply by observing the variation of the contents of the set of rules defining the subjacent device. Returning to the comparison with solid mechanics, we can say that the behavioral evolution of a device may be compared to the velocity of a solid, if we associate the device's set of rules to the position of the solid, while velocity may be deduced from position as its temporal derivative.

The behavioral evolution of a device refers to a local evaluation of the way adaptive devices evolve, and may be formally expressed by establishing some suitable law for the accommodation of the device's rule set in response to the execution of adaptive moves by the device. Consequently, the behavioral evolution of a device may be understood as the law of variation of the subjacent device's recognized language (such law of variation captures, and can somehow be used to quantify the dynamic effect of adaptive actions over the language accepted by the evolving adaptive device).

Additionally, if we define a behavioral evolution growth rate, we can observe that the rate of variation of the behavioral evolution can be compared to the acceleration of a moving solid, which in turn corresponds to the temporal derivative of its velocity. Similarly, the behavioral evolution growth rate for a device may be considered as a local quantification of how fast the adaptive device evolves in its current run. Then, the behavioral evolution growth rate may be measured or estimated by determining some accommodation law for the device's rule set, or estimated by means of some acceptable approximation or non-exact calculation.

5. Concluding remarks

In this work, we tried to document some of the recent progress achieved in investigations about application of adaptive techniques in the resolution of problems from a wide variety of areas. In particular, our focus here has been the class of problems involving modeling and simulation, having as the main target the use of digital computers and the application of adaptivity as the background for modeling complex phenomena. The goal was to propose a sufficiently general technique that should allow the easy specification and construction of efficient and fast simulators for reactive phenomena.

With such premises, a methodology proposal was elaborated, specifying that the most adequate configuration for such class of simulators requires the rule-driven approach for modeling the phenomenon to be simulated as some adaptive device, and a discrete event-driven approach for finally implementing the desired simulator.

Several initial experiments have been successfully assayed, all of them based on previously tested adaptive devices, encouraging us to proceed with this project, in order to develop a framework for the specification, implementation and assay of simulators for several purposes. In addition, a side-product of this work is advancing the theory and creating more techniques for spreading the use of adaptivity for multiple purposes and in diverse fields of interest.

Bibliography

Andersson, J.; de Lemos, R.; Malek, S. & Weyns, D. (2009). Modeling dimensions of self-adaptive software systems. In: Cheng, B.H.C.; de Lemos, R.; Inverardi, P. & Magee, J. (eds.). *Software engineering for self-adaptive systems.* Lecture Notes in Computer Science, 5525. Berlin: Springer, pp. 27-47.

Bedau, M.A. (2003). Artificial life: organization, adaptation and complexity from the bottom up. *Trends in Cognitive Sciences,* 7(11), 2003, 505-512.

Chung, C.A. (2004). *Simulation modeling handbook: a practical approach.* Boca Raton (FL): CRC Press.

Hochbaum, D. (ed.) (1997). *Approximation algorithms for NP-hard problems.* Boston: PWS.

Langton, C.G. (1998). *Artificial life: an overview.* Cambridge (MA): MIT Press.

Lewis, H.R. & Papadimitriou, C.H. (1998). *Elements of the theory of computation.* 2nd ed. Upper Saddle River (NJ): Prentice-Hall.

Neto, J.J. (2002). Adaptive rule-driven devices: general formulation and case study. In: Watson, B.W. & Wood, D. (eds.). *Implementation and application of automata: 6th International Conference, CIAA 2001.* Lecture Notes in Computer Science, 2494. Berlin: Springer, pp. 234-250.

Ross, S.M. (2002). *Probability models for computer science.* San Diego: Harcourt Academic Press.

Tchitchigin, A.; Talanov, M.; Safina, L. & Mazzara, M. (2016). Robot dream. Online: https://arxiv.org/abs/1603.03007 .

Chapter 9

Technology and society: the impacts of the internet of things on individuals' daily life

Mariana Claudia Broens
Department of Philosophy
São Paulo State University (Unesp), Marília, SP, Brazil
mbroens@marilia.unesp.br

João Antonio de Moraes
Department of Philosophy
Campinas State University (Unicamp), Campinas, SP, Brazil
moraesunesp@yahoo.com.br

Arturo Forner-Cordero
Polytechnic School
University of São Paulo (USP), São Paulo, SP, Brazil
aforner@usp.br

Abstract

The main aim of this chapter is to investigate possible social and ethical impacts of the Internet of Things (IoT) in common human action, especially the spontaneous activities of everyday life. As we will argue, the continuous usage of the IoT is changing the way individuals interact with the environment, generating new possibilities of perception/action due to technological affordances. We consider that this new kind of affordance is contributing to a merging of traditional and technological environments, creating hybrid niches. We will illustrate this merging process with two case studies: the emergence of hybrid beings, in the form of individuals who incorporate habits enabled by IoT technologies, and the development of the Internet of Robotic Things (IoRT). Finally, we will discuss possible positive and negative consequences of the increasing usage of IoT technologies.

1. Common spontaneous actions and *affordances*

The present investigation on possible social and ethical impacts, positive and negative, of the new information technologies in human society will be made from the perspective of the ecological theory of perception/action. This theory has as an exponent J.J. Gibson (1986), according to whom the direct perception of "affordances" (without the need for internal information processing) allows common spontaneous everyday actions of embodied embedded agents. From this perspective, agents who share a co-evolutionary history with their ecological niches directly perceive the possibilities of action offered by the various components of these niches.

Gibson (1986, p. 128) argues that: "[...] a niche refers more to how an animal lives than to where it lives". In this sense, the concept of niche goes beyond the mere place where animals live; it is the result of a long-term history of interactions between organisms and the environment. In other words, a niche is formed by the group of features that are appropriate to the organism and that are significant to its kind. For this reason, Gibson considers that the niche can also be conceived as a set of affordances. As stressed by Gibson (1986, p. 127):

> *The affordances of the environment are what it offers the animal, what it provides or furnishes, either for good or ill. [...] [Affordance is] something that refers to both the environment and the animal in a way that no existing term does. It implies the complementarity of the animal and the environment.*

To clarify the concept, Michael Turvey (1992, p. 177) stresses that: "Affordances are dispositional properties of the environment". These are properties of the environment that directly allow the emergence of some other property in certain circumstances. The nature of dispositional properties is potential, that is, they may or may not actualize themselves depending on specific circumstances. Properties such as "catchable", "climbable", "drinkable", and "edible", amongst many others, are examples of dispositional properties because different environmental elements allow the actualization of possibilities of action for specific organisms in specific circumstances. In the case of humans, from a social point of view, they also directly offer varied social affordances to other humans. As Gibson (1986, p. 128) points out:

> *The other animals afford, above all, a rich and complex set of interactions, sexual, predatory, nurturing, fighting, playing, cooperating, and communicating. What other persons afford, comprises the whole realm of social*

> *significance for human beings. We pay the closest attention to the optical and acoustic information that specifies what the other person is, invites, threatens, and does.*

According to the direct perception theory, a large part of human social interactions results from social affordances, directly provided to people by other people, using facial expressions of emotions, gestures, and body language, amongst other means (MacArthur & Baron, 1983). We consider that human actions involving complex moral abilities – such as benevolence, altruism, loyalty, friendship, and collaboration – also result from direct reciprocal interactions, as well as other environmental and social factors. Long-term collective social and moral habits that emerge from direct reciprocal interactions among human agents are fundamental for social balance in human societies. Examples of these collective habits can be found in a varied range of personal and professional everyday activities: basic ones, such as those habits related to the respect between co-workers or how to properly share the common space at work, and personal ones related to our expectations as to how a friend or a sibling is supposed to behave in certain circumstances, among many others.

Furthermore, the shelters humans build in urban contexts, the different tools and instruments they produce, and, more recently, new technological devices, offer a rich diversity of possibilities of action. New information technologies are providing forms of social contact beyond the limits of immediate face-to-face interaction, allowing the establishment of new social habits that emerge from so-called technological affordances, which are "[...] functional, second-order, emergent properties of a perception/action system" (Quilici-Gonzalez et al., 2014). These new technologies, such as the Internet of Things (IoT), are increasingly integrated in human niches, substantially changing previously well established collective habits.

2. IoT and the arising of new kinds of social interaction

The IoT is a computing concept used to describe the network formed by everyday things that are able to connect with each other via the Internet, providing new services to the users and producing a massive amount of data. The novelty of the IoT is that the data are not just created by the users interacting with the devices; they are also created by the devices themselves in a way that is not necessarily controlled by their users, allowing different kinds of communication: human/human, human/thing, and thing/thing. Many different kinds of devices can be interconnected, gathering data and providing services: sensors, actuators, field operation devices, communication devices, and monitors that can measure different phenomena. These are just a few examples.

One of the main features of the conventional Internet is that computers constitute the overwhelming majority of the agents participating in the network. On the other hand, one of the most striking features of IoT is that, in addition to computers, we now have as agents any device/system that processes, collects, or manages information and has an IP address. As pointed out by Gubbi et al. (2013, p. 1645):

> *Ubiquitous sensing enabled by Wireless Sensor Networks (WSN) [...] offers the ability to measure, infer and understand environmental indicators, from delicate ecologies and natural resources to urban environments. The proliferation of these devices in a communicating-actuating network creates the Internet of Things (IoT), wherein sensors and actuators blend seamlessly with the environment around us, and the information is shared across platforms in order to develop a common operating picture (COP).*

Gubbi et al. (2013) probably mean that certain IoT devices can react without direct human intervention. However, quite possibly most of these devices will react in the way that the programmer or engineer decided they would react. For instance, a person may have a sensor/actuator that controls the water level in a drinking fountain for animals. If there is no more water, the sensor will send a message to the person. If the person does not answer the message within a predetermined time, the system can automatically refill the drinking fountain. Although it seems that the system has reacted to the environmental stimuli and has made a suitable decision without the direct intervention of a person, it did what a computer scientist or a programmer had designed it to do in this type of situation. The system has a certain degree of autonomy, but is far from acting independently, in the sense that one considers that a human being is autonomous.

As a result of the new services provided by IoT, new possibilities of human perception/action allowed by technological affordances are creating second order technological niches, in which the IoT is deeply integrated. New social habits related to a huge variety of human personal and professional interactions have progressively emerged in these second order niches. For instance, instead of going to work, people can work at home, interacting with his/her co-workers using Skype, WhatsApp, or other media, enhancing his/her professional capabilities (for example, using translation applications). Personal relationships and friendships can begin by means of social networks such as Facebook or Twitter, almost anything can be purchased online, political activities can be organized and implemented, and the elderly and people with special needs can be monitored and assisted, among many other possibilities. Unlike immediate ecological interactions that occur in traditional human niches, these niches constituted by technological affordances are a product of second order ecological interactions.

The traditional and new niches can no longer be easily differentiated, because they are constantly integrated and interwoven, especially in urban areas of Western culture where ubiquitous computing proliferates. As a result, actions performed in second order niches have a direct impact on the ecological dynamics of the traditional ones, and vice-versa, allowing the emergence of hybrid overlapped niches. Sorj & Fausto (2016, p. 31) point out that:

> *To the extent that the "Internet of Things" advances, the merger between the two [traditional and second order] worlds will be such that it will require new conceptualizations. We live in a historical period in which a world that is coming coexists with another that is leaving, even though its features remain present.*

A current example of the blurred line existing between offline and online human interactions is the "augmented reality" (AR) technology, which offers a considerable amount of technological affordances. AR can be defined as the result of the superposition of online images, audio, and other sense enhancements over traditional human niches, displayed in real time (Cassella, 2009). According to Huang et al. (2013, p. 1):

> *With rapid advances in user interface and mobile computing, comes the ability to create new experience that enhances the way we acquire, interact and display information within the world that surrounds us. Mobile technology improvements in built-in camera, sensors, computational resources and power of cloud-sourced information have made AR possible on mobile devices.*

The application program Pokemon Go illustrates the use of AR for entertainment. This app has been developed for smartphones using AR to allow the players to capture digital creatures that are embedded in the environment according to the smartphone display. In order to play, the user has to move around in the environment, directing the camera of his/her mobile device until a digital creature appears as a part of the landscape displayed on the screen. The following pictures (Figs. 9.1 and 9.2) illustrate the emergence of a hybrid niche.

Figure 9.1: Traditional human niche.

Figure 9.2: Hybrid niche. Images courtesy of Carol Bataier.

Thus, ubiquitous technologies like IoT allow new kinds of second order online interactions in ecological niches, not just mediating the relationship between agents and the environment, but also altering the perceived environment itself and the collective human habits related to it. In the next section, we will provide two examples of possible impacts of IoT upon human everyday life.

3. Implications of the influences of IoT in social life: case studies

3.1 IoT and the emergence of hybrid beings

IoT technologies are promoting deep changes in contemporary society, mainly in collective social habits, due to their increasing presence in the daily life of the individual. Collective human habits are changing mainly as a result of: i) increasing popularization of personal computers and mobile devices, enabling a huge variety of online interactions; ii) the development of ubiquitous computing with global range; and iii) many projects aiming to bring internet connections to remote areas.

The constant proliferation of IoT is creating a growing dependence of its users, mainly because information and communication technologies (ICT) act as mediators between individuals and the environment in at least two different ways:

- *Partially controlled by its user*: as occurs in online banking transactions (home banking), online shopping in virtual stores (e-commerce), apps for personal relationships (e.g., Tinder), television via online broadcasting (SmarTV, Netflix, etc.).
- *Controlled by others* (people or machines): biometric systems for lending books in libraries, online curricula, and tax declarations, among others.

Rafael Capurro (2008) argues that the dependency relationship with ICT has led to individuals acting digitally in the environment, generating new habits. Moreover, the degree of environmental embeddedness of these technologies is such that: "The view of computers as something 'other' is disappearing, i.e., they are less and less 'something' or 'other-than-us' and permeate the world in which we – or, more precisely, some of us – live". In other words, the boundaries between traditional and second order niches, and the individual's unfamiliarity in interacting with IoT technologies, are concomitantly vanishing (for instance, there is nothing unusual in the "autonomous" communication between one's smartphone and smartwatch). Furthermore, the dependency relation between individuals and IoT technologies, and the increasing familiarity with these technologies, culminate in a problematic excess of confidence in the performance of the device. The app called Waze (real-time GPS navigation program) is an example of transferring human perception/action to an online device: the app becomes responsible for "deciding" the route to be adopted, and can change it by itself during the journey, sometimes reaching an incorrect location; in using it, the driver acts in an "automatic way", only following the app's suggestions.

We believe that the ways the IoT is influencing human collective habits are responsible for the emergence of hybrid beings (Moraes & Andrade, 2015). Such beings emerge as a result of their perception/actions in traditional and second order niches over the course of time. As mentioned, the ways that individuals interact in traditional niches constitute well established social habits that can be reproduced, at least partially, in second order niches. However, brand new online social habits may emerge. These new habits, in turn, can offer second order feedback that may alter habits in the traditional niches. Over time, the feedback processes between traditional and online human/human, human/machine, and machine/machine communication may promote the emergence of hybrid beings that interact in hybrid niches in a merged way by means of IoT devices.

One example of individuals who are hybrid beings is the so-called Generation Z, born around 1995 or later. They are known as digital natives, having never known a world without the presence of Google, Wikipedia, Twitter, and Facebook, and for them such companies are not service providers, but verbs. They do not know a world without notebooks, tablets, broadcasting, e-mail, smartphones, smartwatches, touch screens, and so on. It should be noted that hybrid beings go beyond mere avatars, alter egos, or cyborgs, and that the actions of such individuals in hybrid niches, mediated by IoT and by offline structures, become habits, hence constituting part of their own identities.

3.2 The Internet of Robotic Things - IoRT

Recent studies in the field of Robotics have proposed the concept of Internet of Robotic Things (IoRT). Using the online connectivity opportunities provided by conventional IoT, Stuart Carlaw and colleagues state that robots with IoT technologies:

> *[...] use local and distributed "intelligence" to determine a best course of action, and then act to control or manipulate objects in the physical world, and in some cases while physically moving through that world. (ABI Research, 2014, our emphasis)*

Robots are increasingly interacting with humans. An emerging area of robotics research focuses on service robots aimed at providing care, help, assistance, rehabilitation, or training to the human user. These medical robots can be classified in several ways. However, a major distinction can be made between therapeutic and functional assistance robots. Therapeutic robots are those intended to help the application of a therapy, either in a clinical environment or at the home of the patient. In this category can be included robots designed to obtain diagnostic measurements by means of IoT technology.

Due to their nature, therapeutic robots must always be under the direct supervision of a clinician, usually a physiotherapist, and they enable increased intensity of the therapy applied to patients. They also reduce the physical effort of the therapist, while the patient usually feels engaged in the task with the robot, hence improving the rehabilitation process.

Functional assistance robots, in turn, are those designed to support a function lost by the patient. These devices can be robotized prostheses or orthoses, and usually have a strong physical interaction with the patient (for example, they can be implanted). In the design of these robots and their interfaces with humans, there are two main ethical issues that are already well established: safety and privacy.

In the case of robotic prostheses and exoskeletons, there is a strong physical and cognitive interaction with the user. Therefore, in order to comply with the major requirement of safety, it is necessary to include this requirement at the earliest stage of the design process. In a case study concerning the analysis of a modular exoskeleton design methodology, Souza et al. (2016) described a formal procedure to define and refine requirements, from the abstract level to the engineering level.

As with other IoT devices, all assistive robots store and share data obtained from the user. In the case of therapeutic robots, the data may be handled by the clinician, while for functional assistance robots that are outside the clinical environment, the destination of the data is not so clear. The data collected can be directly used to improve the behavior of the robot. For instance, a certain movement pattern (for example, climbing the stairs) can be identified that is specific to the individual. A robotic prosthesis can store and use the data to adapt its behavior and help the user. However, the data can also be used to improve a commercial device, without the knowledge of the user. This issue is commonly addressed in institutional review procedures concerning the data obtained from experiments, and the same protocols could be applied to functional assistance robots.

4. Final comments

It seems reasonable to believe that the dependency relation between individuals and IoT technologies will continue to grow rapidly. It is therefore important to evaluate positive and negative aspects of this dependency relation in terms of the possible ethical consequences of IoT in human society,

On the positive side, IoT technologies contribute to enhancing the performance of individuals engaged in personal and professional activities, especially by providing information and means of communication in a rapid and easy way. However, without being technophobic, the continuous, almost addictive, usage of IoT technologies may lead to negative side effects associated with deep interference in the long-term dynamics of human perception/action.

As mentioned, we consider that a significant amount of social and moral human spontaneous everyday actions is based in the direct perception of social affordances. In turn, social affordances allow the emergence of long-term habits, mainly cooperative, related to altruism, friendship and loyalty, among others, that have succeeded until today to provide basic stability to human societies. However, IoT technologies are rapidly changing this scenario by mediating human-human and human-environment interactions. The Pokemon Go app, for example, provides a good illustration of this new scenario.

There is evidence that the Pokemon Go app is helping to motivate those users with autism, Asperger syndrome, severe depressive conditions, or even a sedentary lifestyle to walk around, perform physical activities, and have at least some social contact, as pointed out by Cao (2016). However, on the negative side, there have been many complaints about the way that Pokemon Go can distract people and cause various kinds of accidents, as well as threaten individual privacy. As Biddle (2016) remarks, in order to play, the app requires the whole attention of its users and an intense and continuous access to their personal information (such as location) and the smartphone camera, microphone, settings, and operating system. It is noteworthy that the terms of use proposed by the developer of the app state that the collected information may be shared with others without the explicit consent of the user.

In turn, many ethical and legal issues arise in the development and deployment of robots, such as IoRT devices (Bogue, 2014). In robotics, ethical considerations must include the ethics of the robot's behavior, the ethics of the designer and user of the robot (including the robot operators), and the ethics of the human-robot interaction. As mentioned, therapeutic robots must always be under the direct supervision of a therapist. However, the accurate data collected by the robots, with sensitive information about the user's health status, may be accessible to others, including health insurance companies, without the user's explicit consent. This situation constitutes a potential threat to the user's own interests.

IoT technologies generate an enormous amount of data, including those concerning aspects of a person's professional and personal life. The most evident problem is that professional and personal data can be easily traced and used without the person's knowledge (in some cases, such an ability may be justified and used to protect collective interests in a democratic society, but this may not always be true). Furthermore, the number of spontaneous actions directly provided by social affordances may gradually diminish as a result of indirect interactions mediated by IoT.

Finally, the information provided by IoT communication processes may directly interfere in well established basic everyday habits (such as paying attention while driving, or respecting the personal privacy of others) that help to preserve social balance and moral conduct, without individuals being aware of it. This interference may occur, for example, in the continuous offering of specific kinds of online services, which may involve undeclared political or financial interests and use the collected data for unknown purposes. For these reasons, ethical considerations of the possible long-term consequences of IoT technologies must be discussed in a society that is committed to mutual respect.1

---

[1] This chapter discusses some of the results of research developed in collaboration with Maria Eunice Quilici Gonzalez, Guiou Kobayashi and José Artur Quilici-Gonzalez.

Bibliography

ABI Research (2014). The internet of robotic things. 20 pp. Available at: https://www.abiresearch.com/market-research/product/1019712-the-internet-of-robotic-things/.

Biddle, S. (2016). Privacy scandal haunts Pokemon Go's CEO. Available at: https://theintercept.com/2016/08/09/privacy-scandal-haunts-pokemon-gos-ceo/.

Bogue, Robert. (2014). Robot ethics and law. Part one: ethics. Part two: law. *Industrial Robot: An International Journal*, 41(4,5), 335-339, 398-402.

Cao, R. (2016). How 'Pokemon Go' is helping kids with autism and Asperger's. Available at: http://edition.cnn.com/2016/08/05/health/pokemon-go-autism-aspergers/.

Capurro, R. (2008). Interpreting the digital human. In: Buchanan, E. & Hansen, C. (eds.). *Proceedings – Thinking critically: alternative methods and perspectives in library and information studies.* Milwaukee: Center for Information Policy Research, University of Wisconsin, pp. 190-220. Modified version available at: http://www.capurro.de/digitalhermeneutics.html.

Cassella, D. (2009). What is augmented reality (AR): augmented reality defined, iPhone augmented reality apps and games and more. Available at: http://www.digitaltrends.com/features/what-is-augmented-reality-iphone-apps-games-flash-yelp-android-ar-software-and-more/.

Gibson, J.J. (1986). *The ecological approach to visual perception.* Boston: Houghton-Miffin.

Gubbi, J.; Buyya, R.; Marusic, S. & Palaniswami, M. (2013). Internet of Things (IoT): a vision, architectural elements, and future directions. *Future Generation Computer Systems*, 29(7), 1645-1660. Available at: http://www.sciencedirect.com/science/article/pii/S0167739X13000241.

Huang, Z.; Hui, P.; Peylo, C. & Chatzopoulos, D. (2013). Mobile augmented reality survey: a bottom-up approach. 35 pp. Available at: http://repository.ust.hk/ir/bitstream/1783.1-8184/1/v2acmsmallsample2.pdf.

MacArthur, L.Z. & Baron, R.M. (1983). Toward an ecological theory of social perception. Psychological Review, 90(3), 215-238.

Moraes, J.A. & Andrade, E.B. (2015). Who are the citizens of the digital citizenship? *International Review of Information Ethics*, 23(11), 4-19.

Quilici-Gonzalez, J.A.; Broens, M.C.; Gonzalez, M.E.Q. & Kobayashi, G. (2014). Complexity and information technologies: an ethical inquiry into human autonomous action. *Scientiae Studia*, 12 (special issue), 161-179.

Sorj, B. & Fausto, S. (2016). Prefacio. In: Sorj, B. & Fausto, S. (eds.). *Activismo politico en tiempos de internet.* São Paulo: Edições Plataforma Democrática.

Souza, R.S.; Sanfilippo, F.; Silva, J.R. & Forner-Cordero, A. (2016). Modular exoskeleton design: requirement engineering with KAOS. *6th IEEE International Conference on Biomedical Robotics and Biomechatronics (BioRob).* Singapore: IEEE, pp. 978-983.

Turvey, M.T. (1992). Affordances and prospective control: an outline of the ontology. *Ecological Psychology*, 4, 171-187.

Chapter 10

The revisionist strategy in cognitive science

Samuel C. Bellini-Leite
Department of Philosophy
Federal University of Minas Gerais, Belo Horizonte, MG, Brazil
samuelcblpsi@gmail.com

Abstract

Cognitive science has recently been facing a breakdown into various distinct approaches, such as the connectionist approach, the embodied approach, the dynamical systems approach, probabilistic approaches, among others and also a relative detachment of disciplines. Much of this happened after theoretical issues related to the frame problem. This work provides an analysis of the frame problem in order to propose the revisionist strategy for cognitive science. Such a methodological strategy aims to describe a recipe for how cognitive science should proceed in dealing with the frame problem. The idea is that theoreticians should start by identifying various sub-instances of the frame problem (of which we provide a first list) to be used as guides for reformulating frameworks of cognitive science and how they relate. It is probably the case that a diverse range of approaches will be needed to deal with different sub-instances. So, it might be the case that diverse general cognitive approaches will be needed for a computationally feasible account of the mind.

1. Introduction[1]

When the foundations of cognitive science were conceived in the 1960's, a clear, unified and coherent concept of the mind was established. It was proposed that a mind was an information processing mechanism that performed serial manipulations on symbols. Symbols were said to represent or refer to objects in any environment and that processing them could guide the actions of agents who possessed such computing abilities. This classical computational theory of mind united various disciplines such as artificial intelligence, philosophy, psychology

---

[1] This paper won first place in the Marcelo Dascal Prize on Cognitive Science of 2015, for student essays.

and linguistics in the common goal of understanding mind as a computation. However, since then various practical and theoretical difficulties emerged as a consequence of this foundational thesis. The most notable of these difficulties is called the *frame problem*. Such problem can be summarized as how an information processing system can relevantly deal with a changing environment. Now faced with these difficulties, cognitive scientists are either reluctant to define their foundations precisely (e.g. the nature of computation which is used by a cognitive agent) or have changed them in some radical fashion (e.g. radical embodiment). This also caused a breakdown of cognitive science into various distinct approaches, such as the connectionist approach, the embodied approach, the dynamical systems approach, probabilistic approaches, among others, and also a relative separation of disciplines. Also, there are those who have argued that the frame problem indicates that the mind is not an information processing mechanism (such as Dreyfus, 1972, and perhaps Fodor, 2001), that is, that there is no hope in solving the problem within cognitive science.

I propose in this paper to describe the revisionist strategy, which is not a solution but rather a recipe for taking the frame problem seriously. In a nutshell, the strategy is to identify sub-instances of the frame problem and constrain psychological and neuroscientific theories of cognition according to such sub-instances. Thus, I discuss the frame problem in detail to show why the revisionist strategy is necessary and identify a list of sub-instances of the problem which must necessarily constrain theories if they are to remain in the realm of cognitive science.

2. The frame problem

The frame problem emerged as a result of a methodology in Artificial Intelligence (AI) which is retrospectively called Good Old-Fashioned Artificial Intelligence (GOFAI, see Boden, 2014, for a recent review). The project of GOFAI was to make the computer agent represent the world by means of formal symbols which stand for aspects of the world in an internal model. In such approach, by means of an external programmer, instructions are given on how to manipulate these formal representations. These symbols are purely formal, syntactical in nature, however they are matched with meaningful stimuli for the outside observer.

McCarthy & Hayes (1969), artificial intelligence scientists following the symbolic agenda, started to model their AI by naming the complete state of the universe at a given time a "situation". However, given that the universe is too large, only a few facts about the situation was to be represented, instead of the whole situation. These facts were used to predict further facts about the current situation or about future situations. It is understood that an action changes a current situation, however, it does not change all elements of the situation, since there are things that change and things that do not change given the action. A problem arose in the task of determining how an agent can determine, in a situation, what should change and what should not, given a specific action. Originally, they had to type in an incredibly huge list of statements, called frame axioms, determining what would not change in result of an action. McCarthy & Hayes (1969) saw this issue as one amongst many other technical difficulties they had discovered and coined the term "frame problem" to refer to this specific difficulty. However, this technical difficulty turned out to be serious enough to require extensive debates between philosophers and artificial intelligence scientists (see Pylyshyn, 1987).

Although there was some consensus in debates over how to understand the frame problem (Pylyshyn, 1987), various philosophers and scientists contributed with unique ways of defining and understanding it. One of the biggest disagreements was whether the frame problem was a specific problem of the field or if it reflected the manifestation of a more general problem. The specific problem may be called the "frame problem proper" (a use suggested by Dennett, 1987). To illustrate it, Dennett (1987) imagines a robot whose task is to go into a room and get an extra battery from a wagon that has a time bomb attached to it. In its first attempt, it pulls the wagon out of the room but explodes, since it brings the bomb along. So, its designers had to teach it how to plan considering the side effects of its actions (considering changes and non-changes of the situation after its action). However, since it had to consider so many non-changes, such as that the color of the room will not change, that the wagon will not change its shape, that the battery will not change any of its physical properties either, and that the wagon will not break by being pulled, the bomb exploded before it could act. Hayes (1987) argues this is the central point of the frame problem. That since these things could in principle change, these frame axioms were needed in order to assert that in that specific time, in that situation, there would be no changes.

> *In this ontology, whenever something MIGHT change from one moment to another, we have to find some way of stating that it DOESN'T change whenever ANYTHING changes [...] That is the frame problem. If there are 100 actions, and 500 labile properties and relations, then we might need 50,000 of these silly "frame axioms" [...]* (Hayes, 1987, p. 125).

As Janlert (1987), Haugeland (1987) and Dreyfus (1972) notice, this seems to be a problem that arises from representing the world in such quasi-linguistic form, considering situations as the state of the universe and dealing with knowledge in encyclopedic statements. Since getting rid of the consideration of non-changes raises other related problems as well, philosophers like, Dennett, Haugeland, Dreyfus, and Fodor suggested that the frame problem proper was only an instance of a broader problem, which they call the "general frame problem".

3. The general frame problem and the revisionist strategy

Janlert (1987) argues the original frame problem is just one manifestation of a general problem of dealing with the phenomenon of change. His general definition is the following: "The general frame problem is the problem of finding a representational form permitting a changing, complex world to be efficiently and adequately represented" (Janlert, 1987, p. 7).

Dennett (1987) defines it somewhat differently. He suggests the problem is one of making good use of the knowledge one has given the requirements of the time-pressured world. He suggests

> *the frame problem [...] concerns how to represent (so it can be used) all that hard-won empirical information [...] a problem that arises independently of the truth value, probability, warranted assertability, or subjective certainty of any of it. Even if you have excellent knowledge (and not mere belief) about the changing world, how can this knowledge be represented so that it can be efficaciously brought to bear?* (Dennett, 1987, p. 52)

He argues the agent that solves the frame problem ignores most of what he knows and works with the portion of his knowledge that is relevant to the situation at hand. Also, he must know how to choose effortlessly the portion of knowledge he is to work on.

Glymour (1987, p. 65) also agrees that there is a general problem. "I think the 'frame problem' is not one problem, but an endless group of problems concerned with how to characterize what is relevant in knowledge, action, planning, etc." His main contribution, however, is seeing the problem by means of its relation to the computability constraint. He argues that the general problem is one of finding feasible algorithms. This is an important notion which we will have to describe in some detail.

An algorithm can be evaluated by the number of steps it takes or the time it takes to solve a problem, that is, the resources it uses. An instance of a problem is one specific input string for a computational problem. The longer (the size of) the instance, the greater the resource consumption will be. A feasible algorithm is one in which the worst case resource consumption runs in less than polynomial time. That is, if a polynomial function of the input size is at least as large as the number of steps taken by the algorithm to solve the problem in the worst case scenario, then the algorithm is feasible. In other words, given that a problem with the size X requires time T, the polynomial function with the input of size X needs to result in a number larger than T. These are problems of the class P, those whose solution can be found in polynomial time.

In order to understand the feasibility of algorithms, computer theorists have created the class of problems NP, which can be defined as the class of algorithms for which the solution can be checked to be correct in polynomial time. Therefore, this class includes the problems of class P, but also problems for which there are no known algorithms to solve it in polynomial time. Notice that a clear distinction is made here between the time taken to solve a problem and the time taken to verify whether a solution is correct.

The class of NP-hard problems is composed of reference problems that are equivalent to any NP problem. That is, any problem of class NP can be converted into a problem of class NP-hard by an algorithm that runs in polynomial time. Therefore, if a polynomial algorithm is found for a problem of class NP-hard, every problem of class NP would be solvable in polynomial time. This would imply that the class P is the same as the class NP. However, computer theorists believe this not to be the case, although no one has been able to give a definitive answer to this question to date. Among the class of NP-hard problems are the NP-complete, which are reference problems that are equivalent to any NP problem, which also have the property that its solution can be checked in polynomial time. Therefore NP-complete problems are in the intersection of the NP and NP-hard problems. (These technical concepts can be found in Ladner, 1975, and in Cormen et al., 2001).

To solve the frame problem as analyzed by Glymour (1987), therefore, one would need to follow the computability constraint. There are two strategies for that, what I call the 'the revisionist strategy' and what I term 'the magic strategy'. When NP-hard problems are found, the revisionist strategy proposes that the theorist should revise how it is assumed that cognitive agents compute (see van Rooij, 2008, 2012). The magic strategy proposes that for cognition to work in complex environments one would need to be able to find a solution to NP-hard problems, which is beyond the current capability of computation theorists and mathematicians. So, this is a fundamental difference, while the revisionist strategy attempts to *avoid* NP-hard problems in theorizing about cognition, the magic strategy believes we would only be able to fully theorize about cognition if NP-hard problems could be *solved*.

It would seem that no one would advance the magic strategy. However, it seems Fodor (1987, 2001), who is one of the main advocates of the thesis that there is a general frame problem responsible for other specific ones, has something of the sort in mind. He seems to affirm implicitly that human brains solve NP-hard problems while mathematicians and computer theorists do not, and for that reason he sees a grim future for cognitive science. That is not, precisely, how he puts it however. Fodor (1983) believes the mind is divided into two general types of structures, a modular structure and a general (not modular) system. The main characteristic of modules used for his argument on the frame problem is 'information encapsulation'. By this he means "[...] one that has access, in the course of its computations, to less than all of the information at the disposal of the organism whose cognitive faculty it is" (Fodor, 1987, p. 139). In contrast, the mechanisms that deal with relevant thinking in complex situations are not encapsulated. That is, any knowledge that could be used must be available, the mechanisms need to be sensible to every aspect of our webs of belief (Fodor, 1983, 2001). That is why we interpret Fodor (1987) as following the magic strategy, since this necessarily invokes the need for a solution of NP-hard problems. Either that or it is not in the realm of cognitive science at all (Dreyfus, 1972; Fodor, 2001). Particularly, Fodor (1987, 2001) is skeptic that computations (a deduction process) could deal with the necessary inferences for cognition in context, induction and abduction.

In contrast, we get hints of what the models of cognitive science should keep, avoid or revise by following the computational reading of the frame problem proposed by Glymour (1987). For example, he presents one such hint: "All known algorithms for testing for consistency are worst-case exponential. If we decide nonetheless to do consistency tests, the machine must do as few as possible" (Glymour, 1987, p. 71). This is what inspired the current paper. This sort of hint states that algorithmic problems indicate problems that should be revised at the computational-level (or properly psychological) theory and suggests revisions (see van Rooij, 2008, 2012). This paper proposes an expansion or an elaboration of such idea. The revisionist strategy, as expanded, is to use not only algorithmic constraints but *various types of sub-instances of the frame problem* as guides to provide hints for how psychological models of cognition are to be pruned. Thus, if any part of a psychological theory of cognition falls into any sub-instance of the frame problem, it should be revised and pruned; the frame problem should not be ignored but used as a guide for feasibility of cognitive theories.

4. Sub-instances of the general frame problem

Hayes (1987) argues that the frame problem proper is the only real frame problem, that there is no sense in speaking of a general frame problem. He believes we cannot confuse the frame problem proper with other related problems because that would be a loss of scientific progress. He argues that there is a generalized artificial intelligence problem (GAIP), which is how to get a machine to reason sensibly about the world. He blames philosophers for equating the frame problem with the GAIP, when it is rather a sub-aspect of the GAIP. This is problematic, so the argument goes, because AI scientists worked hard to identify specific sub-problems of GAIP, one of which is the frame problem proper, and the progress was in having carefully distinguished these so they could be tackled one by one, as the scientific process requires. But what Hayes overlooks is that there are various difficulties which do not feature in the general frame problem, such as learning, translating, long-term planning, abstract problem solving, etc. The general frame problem is not the whole problem of artificial intelligence (it is not the GAIP).

I do, however, sympathize in part with both sides. I agree that subdividing the frame problem helps us understand it better. Specifically to our goal, it helps us study how cognitive science can avoid each of these sub-instances of the problem. However, we agree with philosophers that there is a broader frame problem that unites all these sub-instances, and particularly that solving one sub-instance alone is not going to help cognitive science progress, if it just brings up new related problems. I will now list a few sub-instances of the problem so they can serve as steps for the revisionist strategy. One can see these sub-instances as components that form a single problem.

The first obvious sub-instance is the frame problem proper itself (SI-1) the problem of modelling non-change; or a bit more broadly, the problem of ignoring irrelevant information. Janlert's (1987) definition can come in as a sub-instance as well; we reach it if we extend SI-1 further to include generally the problem of modeling change (SI-2) (what changes along with what does not). Janlert (1987) helps divide some of the other sub-instances (or what he calls related problems) of the general frame problem. He starts by characterizing what is termed the prediction problem (SI-3). He explains that cognitive systems need to have some sort of predicting mechanism, and that these predicting systems need to have information about the current state of the world. Traditionally, in an attempt to deal with the prediction problem (among others), problem worlds were reduced to idealized worlds, worlds with simple shapes such as cubes and pyramids. Obviously, the success of prediction for idealized worlds does not transfer to the real world. Whenever a system starts trying to predict the real world using GOFAI, the reliability of prediction models necessarily decreases. An agent faced with the task of predicting the real world has to have a way of knowing about the world in advance without computational explosions. Time is also a problem. One could end up with a fast but unreliable prediction machine or a machine whose prediction comes after the fact has already happened. The upshot is that a representation does not help in predicting both if it is too simple or too complex.

Janlert explains (1987) that in idealized box worlds the prediction problem does not come up while the frame problem still does, and that such is a simple way of showing how they are distinct. However, the reason philosophers probably would think this as a sub-instance of the general frame problem is because it is an instance of the difficulties of relevant use of information in real space and time. As a response to Janlert, it could be argued that if one were to solve the general frame problem in real world situations, the prediction problem would truly be solved along with it.

Janlert (1987) goes on to explain the bookkeeping problem (SI-4). It is the problem of keeping track of the various plans and hypothetical situations that the agent needs in order to solve a given problem. It includes keeping in storage different descriptions of the situation, maintaining for use what is necessary, and relating descriptions with each other, knowing that one might be the consequence of the other. While this is not the problem of modelling non-change, it is also an instance of difficulties relating to the use of relevant information, therefore, a sub-instance of the general frame problem.

Finally, Janlert (1987) distinguishes the qualification problem (SI-5). The qualification problem is the problem of identifying, in real-world situations, objects that are represented in the agent's mind. We all know what a chair is, but what exactly counts as a chair? Is a tree stump a chair? Does a chair need to have legs? Humans sometimes fail to find the adequacy of real world objects with their own concepts; for an artificial intelligence it is even harder. We seem to use some sort of common sense to be efficient at answering these qualification questions. I believe this is a sub-instance of the frame problem, specifically because agents need to qualify objects as they show up in real time, and not a priori, therefore, it is a problem of relevant use of information.

An important sub-instance that is much related to the general frame problem is the problem of time pressure (SI-6). Dennett (1987) argues that knowledge use must be constrained by the relevant time scale imposed by the world. He makes an important distinction between problems of knowing and problems of using knowledge; induction is for knowing and learning, while all these sub-instances of the frame problem are related to the correct *use* of knowledge. This seems to be a more precise analysis than that of Fodor's (1983, 1987), in which the problem of induction is the same as the frame problem. Therefore, even if the agent had figured out induction and had all knowledge available, he would still have the frame problem if he did not know how to use the knowledge. Fodor (2001) himself stopped talking about induction, switching to abduction in later writings.

Haugeland (1987) speaks of the problem of keeping temporal knowledge up to date (SI-7). Hayes (1987) terms this the updating problem and argues that such is not the frame problem, since the frame problem is a matter of beliefs about a changing world, not about changing beliefs about a world. I side with Haugeland in this, since it is also a consequence of not knowing how to make relevant use of information. If an agent cannot update his belief against change, then he will not be able to use all the relevant necessary information and other sub-instances of the frame problem will invariably follow.

Hayes (1987) argues that the search problem (SI-8) is not related to the frame problem. I agree it is not the same as the frame problem proper, however, it is clearly a sub-instance of the general frame problem. Given that the agent has stored much knowledge, the problem becomes how it is to search through such content efficiently in order to use only the relevant knowledge for the task at hand. Dennett (1987) and Dreyfus & Dreyfus (1987) agree this is a central aspect of the frame problem.

Fodor (1987, 2001) and Glymour (1987) worry about how an agent could reach at least some internal consistency (SI-9). Given that the agent must know so much, it becomes hard to understand how all this knowledge could be at least partially consistent without running into computational problems. As also explained by Cherniak (1986), for relevant use of information at least some of our beliefs need to be consistent, and checking for their consistency must be done just right. Too much checking and we would not have an answer, and too little checking would render the agent irrational. This is deeply related to Glymour's (1987) main worry, the problem of using feasible algorithms (SI-10). It seems almost necessary that for information to be used relevantly by a mechanical agent, it must use only polynomial feasible algorithms, and avoid NP-hard ones. A failure to avoid NP-hard problems would render the computation impossible.

Wheeler (2008), distinguishes two further sub-instances of the frame problem. The intra-contextual frame problem (SI-11), which concerns the problem of how a mechanical agent can use information in a given context to act fluidly and flexibly. Interestingly, this illustrates how 'using information relevantly' does not mean only stored information in the brain but also information one might capture in the environment. Some autistic individuals, for instance, tend to focus too much on irrelevant aspects, for example, ignoring people's faces and focusing on the shape of corners of tables. The second sub-instance is the inter-contextual frame problem (SI-12), the problem of how mechanical agents can deal with multiple contexts, apply knowledge from one context to the other, and learn to act in new contexts. That is, transferring information relevantly in between types of environments.

Finally, although Hayes (1987) argues that the frame problem is not just a general representation problem, if Dreyfus (1987) is correct, the frame problem follows from the attempt of representing knowledge in language-like structures (SI-13). In fact, many of these subdivisions are necessarily a difficulty of making use of content stored in propositions.

Name of sub-instance	Brief description
SI-1. The problem of modelling non - change	How to know what will vary, given an action
SI-2. The problem of modeling change	How to monitor what does change
SI-3. The prediction problem	How to predict in a changing environment
SI-4. The bookkeeping problem	How to keep track of goals
SI-5. The qualification problem	How to identify objects using concepts
SI-6. The problem of time pressure	How to use knowledge effectively
SI-7. The updating problem	How to keep knowledge updated in relation to the changing world
SI-8. The search problem	How to search through stored knowledge effectively
SI-9. The problem of internal consistency	How to keep beliefs coherent
SI-10. The problem of using feasible algorithms	How to avoid NP-Hard algorithms
SI-11. The intra-contextual frame problem	How to deal with the present environment
SI-12. The inter-contextual frame problem	How to transfer knowledge between different situations
SI-13. The problem of non-propositional thought	How non-linguistic thought could work

Table 10.1: List of sub-instances of the frame problem.

Of course, these 13 problems are definitely not the only way to divide the frame problem, it is very likely that various many other sub-instances could be identified. The point is that they do need to be identified and, most relevantly, *used* in order to prune models of cognition. It seems the revisionist strategy is the only one which takes these problems seriously. The magic strategy ignores this sort of detailed attempt to solve problems and hopes some other fundamental principle is discovered which will help us understand relevance and abduction, allowing one to deal with all these issues at once, possibly without invoking computation or information processing. That is, since this fundamental principle would solve relevance and abduction for good, all these sub-instances of the frame problem would also be gone. Another common path is to ignore the frame problem. The revisionist strategy needs urgently to be put in practice in psychology and cognitive neuroscience. To be clear, the point of the revisionist strategy is not to limit our minds to the level of current machines, it is rather to avoid use of descriptions of information processing strategies which machines have already shown to be limited. More generally, any well described problem found in AI and robotics should not be ignored by other disciplines from cognitive science, but used to prune them.

5. Considerations on the revisionist strategy

Psychologists, neuroscientists and other scientists proposing computational-level theories usually take for granted that any computation may be feasible in principle, that is, they do not worry about the failures of real application of theories in artificial intelligence. For instance, a psychologist might propose a propositional thought system which works contextually in real-time. In relation to psychological evidence and applications this might seem fine, but artificial intelligence has already shown that this cannot be feasible in real information processing systems. This is a major mistake if cognitive science is to be practically applied as an interdisciplinary science. The revisionist strategy proposes that using sub-instances of the frame problem as a guide will help alert scientists in relation to their theory's feasibility.

It is also probably the case that a diverse range of approaches will be needed to deal with different sub-instances. For instance, it seems plausible that connectionist strategies will be more suited to answer the search problem (SI-8), the problem of internal consistency (SI-9) and possibly the inter-contextual frame problem (SI-12), while it may be the case that a situated approach (see Clark, 1997) is more suitable for dealing with the problem of time pressure (SI-6) and with the intra-contextual frame problem (SI-11). Also, a predictive coding approach could have more to say on the prediction problem (SI-3) and the problem of non-propositional thought (SI-13). So it may be that diverse general cognitive approaches might be needed for a computationally feasible account of the mind. As Boden (2014, p. 96) has put it: "The motto of both psychological and technological AI today might be 'Let a hundred flowers blossom – or anyway, four.' Those four are the approaches of GOFAI, connectionism, situated AI, and dynamical systems". I would add computational neuroscience approaches such as predictive coding and applications in cognitive developmental robotics (see Asada et al., 2001; and Clark, 2013, 2016).

In reality this is what most sciences of the 21st century look like. For instance, Physics no longer has a unified view of the universe, but rather different theories for each aspect of the universe (such as classical, relativity and quantum). The main difference is that usually each approach in cognitive science attempts to undermine the other. I think this is, for most part, a mistake. The revisionist strategy can be used as a way to help see that differences in models of cognition might be attributed to differences in forms of processing of the same mind, just as there are (incompatible) models of different aspects of the physical world. But when asked: What unites cognitive science? How does it understand the mind? We can still have a very general but good response. Cognitive science is a collection of disciplines studying the mind as a functionally explainable process, as inspired by information processing. Of course, this will exclude from the domain of cognitive science some of the theories that claim that the mind cannot be 'functional' by nature, or that the mind is nothing like information processing; but yes, these views should be excluded since such claims are intrinsically at odds with the foundations of cognitive science.

Bibliography

Asada, M.; MacDorman, K.; Ishiguro, H. & Kuniyoshi, Y. (2001). Cognitive developmental robotics as a new paradigm for the design of humanoid robots. *Robotics and Autonomous Systems*, 37, 185-193.

Boden, M. (2014). GOFAI. In Frankish, K. & Ramsey, W. (eds.). *The Cambridge handbook of artificial intelligence.* Cambridge (GB): Cambridge University Press, pp. 89-107.

Cherniak, C. (1986). *Minimal rationality.* Cambridge (MA): MIT Press.

Clark, A. (1997). *Being there: putting brain, body and world together again.* Cambridge (MA): MIT Press.

——— (2013). Whatever next? Predictive brains, situated agents, and the future of cognitive science. *Behavioral and Brain Sciences*, 36(3), 181-204.

——— (2016) *Surfing uncertainty: prediction, action, and the embodied mind.* Oxford: Oxford University Press.

Cormen, T.; Leiserson, C.; Rivest, R. & Stein, C. (2001). *Introduction to algorithms.* New York: McGraw-Hill.

Dennett, D. (1987). Cognitive wheels: the frame problem of AI. In: Pylyshyn (ed.) (1987), op. cit., pp. 41-64.

Dreyfus, H.L. (1972). *What computers can't do: a critique of artificial reason.* New York: Harper & Row.

Dreyfus, H.L. & Dreyfus, S.E. (1987). How to stop worrying about the frame problem even though it's computationally insoluble. In: Pylyshyn (ed.) (1987), op. cit., pp. 95-111.

Fodor, J. (1983). *The modularity of mind.* Cambridge (MA): MIT Press.

——— (1987). Modules, frames, fridgeons, sleeping dogs, and the music of the spheres. In: Pylyshyn (ed.) (1987), op. cit., pp. 139-149.

——— (2001). *The mind doesn't work that way.* Cambridge (MA): MIT Press.

Glymour, C. (1987). Android epistemology and the frame problem: comments on Dennett's "Cognitive wheels". In: Pylyshyn (ed.) (1987), op. cit., pp. 65-75.

Hayes, P. (1987). What the frame problem is and isn't. In: Pylyshyn (ed.) (1987), op. cit., pp. 123-137.

Janlert, L.-E. (1987). Modeling change – the frame problem. In: Pylyshyn (ed.) (1987), op. cit., pp. 1-40.

Ladner, R. (1975). On the structure of polynomial time reducibility. *Journal of the Association for Computing Machinery*, 22, 155-171.

McCarthy, J. & Hayes, P.J. (1969). Some philosophical problems from the stand-point of artificial intelligence. In: Meltzer, B. & Michie, D. (eds.). *Machine intelligence 4.* Edinburgh: Edinburgh University Press, pp. 463-502.

Pylyshyn, Z.W. (ed.). (1987). *The robot's dilemma: the frame problem in artificial intelligence.* Norwood (NJ): Ablex.

van Rooij, I. (2008). The tractable cognition thesis. *Cognitive Science*, 32, 939-984.

——— (2012). Intractability and the use of heuristics in psychological explanations. *Synthese*, 187, 471-487.

Wheeler, M. (2008). Cognition in context: phenomenology, situated robotics and the frame problem. *International Journal of Philosophical Studies*, 16(3), 323-49.

Chapter 11

Information, context and structure in cognition

Joao Eduardo Kogler Junior
Polytechnic School
University of São Paulo (USP), São Paulo, SP, Brazil
kogler@lsi.usp.br

Paulo Eduardo Santos
Department of Electrical Engineering
FEI University Center, São Bernardo do Campo, SP, Brazil
psantos@fei.edu.br

Abstract

In this essay, we discuss some broad questions about theoretical aspects of cognition, to understand what makes a system of processes a cognitive one. We address the roles of information, context and structural features of the underlying processes, looking for some characteristics for a suitable mathematical structure underpinning cognition. We compare three kinds of agents concerning the information-theoretic aspects of action control: the reactive, the perceptive and the cognitive agent. Then we pave the way for comparing the essential features that distinguishes perceptual and cognitive processes, which lead to the geometric characterization of each one under a view based on Klein's Erlangen Program, in the search for answers to the evolutionary question of how cognition may have emerged. Finally, we consider formal (logic) systems in order to define the semantics of space cognition.

Cognition is a feature presented by a particular kind of agent, which we call a cognitive agent, and our fundamental question is about what makes an agent a cognitive one, independently of being biological, artificial or even abstract. In a traditional view, cognition is a mental activity, typically a human feature, and recently considered an attribute of primates in general, as pointed out by Tomasello & Call (1997). Nevertheless, we do not want to exclude the possibility that other species could exhibit cognition (de Waal, 2016), as well as the eventuality that, at least in principle, even machines could produce artificial cognition (Vernon, 2014).

We started the discussion talking about a generic agent, and now we will focus on its essential feature: an agent is an entity that produces actions, and a history of these actions constitutes a sample of its behavior. The cognitive agent uses cognition to control its behavior, but since we aimed at understanding what cognition is, we will start with a simpler kind of agent, that does not make use of such resource. Another way of controlling behavior is to rely just on perception and this may not require cognition. In addition, a yet more primitive way of producing simple forms of behavior is by reaction.

1. The reactive agent

A reaction is a response that follows a fixed rule, specific to a certain kind of stimulus incoming from the environment surrounding the agent. It is triggered by the detection of the specific stimulus and usually doesn't need more sensorial feedback to proceed, following to the automatic generation of the action. The reactive response is very efficient in the specific situations for which it is tuned; nevertheless, each reaction is not capable to cope with very different situations. Therefore, the purely reactive agent must have a repertoire of reactions, each one featuring a particular kind of action triggered by a restricted class of stimuli. Consequently, purely reactive agents have a limited fitness to few varieties of situations.

2. The perceptive agent

A way to achieve better fitness to a wider spectrum of situations is to produce more complex actions compositionally, by prescribing a set of elementary actions selected from a repertoire of automatic reactions, each one triggered by an event internally generated, following a specific order of execution. Notice that each action executed by an agent changes the state of the neighborhood surrounding the agent. This entails a causation that follows the direction from the agent to the world (Jeannerod, 2006). Therefore, to prescribe an intended set of actions implies in making decisions that involve the prediction of future states of the environment, including anticipating the relations between the agent itself and the surrounding external neighborhood. This scenario is exactly what will put into perspective the differences between cognition and perception.

We refer to a context of an action as the set of relations between two configurations: that of the items in the neighborhood of an agent affected by the action and the agent's own configuration. For instance, the mechanical context in an act of grasping can involve the configuration of objects surrounding the grasped part, as well as the configuration of the body parts of the agent, that could eventually exert some influence in the execution of the motion, besides its own hand.

The prescription of actions suited for some situation relies on the evaluation of their effectiveness, as a criterion to select the more adequate set of elementary actions that will compose the complex action[1]. This evaluation requires a prediction of the context of this set of actions after its execution. A way of predicting the future context is to estimate it from the current one. To do so, the agent rests on its sensorial resources to gather the available information to produce a descriptive representation of the actual state of the local environment. This representation is context dependent, and is generated by the class of processes that corresponds to perception. Thus, we call a perceptive agent one that is able to build representations of the current state of the environment and uses them to estimate the contexts of the actions that it can prescribe to cope with its behavioral situations. It is an entity with better perspectives than the purely reactive agent. Its abilities to cope with situations with increasing complexities grow to the extent that its sensorial system conveys more information to produce better representations of the current environment. Nevertheless, these representations are fitted to the execution of actions and consequently are context dependent, and the amount of internal resources of the agent required for producing these representations grows with the size of the set of situations that the agent would be able to cope with. This addresses the computational complexity of perception, a measure of the estimated size of the set of resources required to meet a given performance.

Consider the case of the visual perception, for example. The amount of resources required to represent the visual information can be translated by the number of neurons and connections comprised by the visual areas of the brain and their sizes. When the environment encompassed in the visual field projects onto the retina, the resulting image decomposes in a number of points corresponding to the resolution of the eye. Each object, surface or region existing in the visual field comprises a description based in the patterns of shape, color, texture, etc. present in those points. To encode these patterns in neural terms requires the decomposition of these aspects in a number of local features detectable by special units composed of neurons, the so-called "receptive fields". In addition, one or more coordinates are required to specify the position, orientation, and direction of motion of each detected feature, and the corresponding receptive field encodes this information. Hubel & Wiesel (2014) proposed a model for the operation of the set of receptive fields in the visual

---

[1] We are not requiring that the prescription of intended actions results from a selection made by a conscious agent, but that it can follow from some impersonal process.

cortex, reflecting the research they did during about twenty-five years, starting about fifty years ago. Their model encompasses a hierarchical architecture with layers of networks made of receptive fields. Each receptive field at a higher level connects to several ones at a lower level. Receptive fields present at the topmost levels in this hierarchy comprise bigger areas of the visual field and the features they detect become more and more complex. By accounting for all these items, we can infer that the amount of resources required for representing the contents of the visual field is extremely high and this becomes yet more evident if one considers the total size of the vision areas comparatively to the total size comprehended by other cortical areas.

Like vision, perception in general requires many resources to represent each configuration spanned in the sensorial field, characterizing a particular state of the environment. The representation of the contexts of each action includes the encoding of the relations between the current configurations, (i) of the environment and (ii) of the agent's own body. Therefore, more than one sensorial modality is involved in the representation provided by perception for each context. All this contextual information provides a sufficiently satisfactory portrait of the agent's current status related to the environment, enabling actions to be executed in very complex situations, with sufficiently accurate prediction of the future state resulting from the completion of the action. Thus, the emergence of perception brought agents with dramatically more sophisticated behavioral repertoires. The perceptive agent presents remarkably higher fitness in comparison with the reactive agent, however at the expense of an exceedingly greater computational complexity.

3. The cognitive agent

The volume of resources demanded by the perceptive agent accrues from the representations of context required to estimate the effects of an action before its execution, in order to enable the prescription of the suitable elementary actions that compose it. The more complex the action, the more sophisticated will be the context, demanding more features for its representation and more time for the estimation process. The reason is clear: the perceptive agent rests on perception, a process that produces representations that rely only on sensory information, which is processed on the fly, in a continuous fashion. This will set limits on the performance and size of the perceptual system, and consequently on the behavioral attainability of the agent.

An evident way of circumventing this scenario is by changing the method, using a new process to produce the representations required by the action system in order to prescribe the suitable scheme of elementary actions to compose a complex action. The expensive item is context dependence. If the representations were made with context-independent items, they could be stored for future use with several related contexts, saving time and freeing computational resources. More precisely, representations should be invariant to context transformations. This addresses a process more sophisticated than perception. This process, cognition, produces knowledge, a context-independent representation.

Therefore, the emergence of the cognitive agent brought a solution that produced enormous flexibility and robustness in terms of fitness to countless environmental situations and varieties, leading to a dramatic sophistication of behavior. The animal kingdom has a number of examples of well-succeeded uses of cognition, besides humans and other primates (de Waal, 2016).

Although the distinction between cognitive and perceptive agents can be clearly stated by the use of context-invariant knowledge by the former, versus context-dependent representations by the latter, it is far from trivial to establish a process-level comparison between cognition and perception. In both cases, the agent needs contextual information in order to make predictions of the results of an intended action before its execution. The success of the task subsumed by the action remounts to the structure of the contextual relation between agent and environment. As action progresses, the behavioral task reaches a successful status that is accessed through an analysis of the sequence of changes in the context underlying that action. We must keep in mind that the context reflects relations, involving the agent and the environment circumscribed by the action, and that these relations change with the progress of the action. The success in the result of the execution of the action will lead this agent-environment relation to a specific situation that should be predicted accurately assuring that the agent has done its part, by acting properly. Thus, in any case, a context representation would be required. Therefore, the cognitive agent must synthesize the action contexts using knowledge about objects and about their relations. This leads to another cognitive feature that is simulation. Cognition is useful for agents that are able to simulate situations of knowledge use, which introduces an extra difficulty in understanding the processes underlying cognition: besides building knowledge, it must use knowledge. Therefore, to emerge, cognition should have encompassed both analysis and synthesis from the beginning, or it would not have brought any evolutionary advantage for animals that started exhibiting it. So, how to conceive such a process?

4. Information geometry of cognition and perception

Before presenting our answer, let us first consider the illustrative case of visual perception and cognition. Goodale & Milner (1992) proposed that there are two vision systems, at least in primates, including humans, one following a dorsal pathway in the brain cortex and the other a ventral pathway. The dorsal vision system starts in the occipital cortex region called V1 in primates and extends dorsally, ending in the motor cortex at the parietal lobe of the brain. The ventral vision system starts also in V1 and extends ventrally to the infero-temporal region. Mishkin & Ungerleider (1982) previously proposed that these two pathways dealt each one with distinct aspects of visual information, attributing to the dorsal system the vision of "where", and to the ventral system the vision of "what". They proposed that actually the two vision systems were completely segregated from each other, from the retina to the parietal or the temporal cortex (Johnson et al., 2008), with the dorsal system providing information about spatial localization and orientation of objects, whereas the ventral system deals with object identification and recognition. Goodale & Milner (1992) argued that the two streams actually started in cortical region V1, sharing common processing of visual information at the initial steps since the retina, and then, after separation following V1, the dorsal stream would be responsible for perception for action and the ventral stream for pattern recognition and object identification. Further research has provided evidence that the two streams are not completely separated, although they seem to deal with different aspects of visual information representation. As argued by Jeannerod (2006), the dorsal stream provides a complete body-centered representation, built with contextual dependence intrinsic to sensory input, and suited for translation into the motor cortex encoding. On the other hand, as argued by Jeannerod, the ventral stream provides also a complete representation of visual information, however progressively independent of context and less observer-centered, as one proceeds deeper into the infero-temporal region. Jeannerod called the dorsal stream the pragmatic pathway, and the ventral stream the semantic pathway, emphasizing a dichotomy due to the differences in context dependence between the representations produced by the two pathways. This view agrees completely with what we have been suggesting in the previous two sections. Nevertheless, Jeannerod also argues that both streams share a common functional organization capable of building complete representations, being the dorsal one committed with action representation, while the ventral stresses the conceptual aspects and ontological relationships. Additionally, he points out that there are links also from the ventral stream to the motor cortex in several instances. The information conveyed through these links employ the conceptual ingredients provided by the ventral stream, combined with further steps of cognitive processing taken at the regions ahead. The further elaboration of action plans employ these unconscious representations and may include conscious intentions. While approaching the motor cortex, these high-level plans must translate into a form suitably encoded into action schemas. Schemas are abstractions, probably learned from experience that captures similarities with familiar situations, eliminating the necessity of analysis at a very detailed level, with the generation of stereotyped responses (Cooper, 2016), in a way that resembles the idea of decomposition into elementary actions, as discussed in a previous section. Once reached this level, it is feasible to close the simulation loop with the dorsal stream representations coming through the other path, and putting into confrontation an estimate based on contextual representations of one another, synthesized via an action plan. This illustrates the way by which cognition can include both analysis and synthesis.

The existence of the two vision systems, the pragmatic and the semantic, using the terms suggested by Jeannerod, clearly suggests the idea of the derivation of cognition from perception. Although both pathways bring about perceptual systems, the semantic pathway resembles the idea of cognition, as a process of building knowledge through context invariance of information.

Under the information-theoretic point of view, both cognition and perception are processes operating on information (data) in hierarchical architectures. At each level of the hierarchy, there are processing modules that summarize big sets of data into smaller collections of properties. In the case of the brain, these modules are networks of neurons. The process of transforming a big set of data into a smaller set of properties, or features of data, generates data for the next level in the hierarchy, which by its own is a processing module that provides a new summarized view made of features of features, and so on. Notice that when we talk about data and processing of data, we are not suggesting the manipulation of discrete symbolic data, as found in computers, but a more general class of processing constituted by dynamical systems operating on signals. Besides, the dynamical systems may operate in a stochastic fashion, by statistically estimating features representing sets of data, or if preferred, by inductively inferring abstractions that generalize sets of data. The set of features that summarizes the incoming data constitutes a representation of the data by the features, in the sense that the features can be used to refer to the relevant information brought by a much bigger amount of data. This kind of representation goes a step further beyond the process of transduction from one physical form to another, as the one that takes place in the sensor organs. The successive stages of representation that take place after the sensorial transduction are all at this same level of abstraction. The success of this scheme for extracting information from data occurs because there is a regularity in the family of processes capable of making the suitable inferences on data. This regularity implies that information can be copied, reproduced, equally used in many instances and so it is a kind of building block useful for building organisms and machines. Both cognition and perception are of this same kind of process, and this is a benefit under the constructive point of view. This can confer to brains and machines a regularity easily reproducible along their whole structure, using the same principles, which is a big economy of effort and resources. So, in what aspect do they differ?

Our proposal as answer to this question is in a rough view very simple: both cognition and perception form these hierarchical chains of processing steps, which are apparently identical. However, the extremities of these chains are connected to conceptual and physically distinct types of terminations: the perceptual chain begins in a sensorial organ or device and ends closely to an action generating system, like the motor cortex or an effector drive. Distinct from that, the cognitive chain begins in some internal source of information, like for instance some point of a perceptual chain, and ends in some other sink of information, without any special conceptual characteristic externally imposed. This is exactly the fundamental point: the endpoints of a perceptual chain are usually information dealing devices that have characteristics imposed by the environment – they reach interfaces that separate regions operating on distinct physical rules and for this require some form of transduction and mapping[2].

With cognition, this does not happen: the cognitive chain operates on internal representations of the same physical nature on every point. This dramatically changes the kind of restrictions that the processing chains have to obey. Using geometric terminology, we propose that the perceptual chain is restricted by holonomic constraints[3], while the cognitive chain has a non-holonomic set of constraints. This conjecture has interesting implications for the modeling of the set of processes underlying these networks under the information geometry standpoint. A yet interesting point to explore is why are these differences in constraint settings referred to as having a geometric nature?

Our proposed answer follows a view of geometry brought by Felix Klein ([1872] 1892) in the so-called Erlangen Program. Under this view, a geometry is a space formed by the classes of invariants under certain transformations. Different invariants lead to different geometries. And this fact has implications for our case: the geometry of cognition is one that arises from a non-holonomic set of restrictions propagated through the chain of transformations that composes the whole cognitive structure, while in the case of perception, the constraints are propagated holonomically. This fact will have a great impact on the kind of invariants resulting from these settings, and consequently these invariants will form conceptually different spaces.

In the case of the perceptual agent, we must recall that the perceptual representations originally provided contextual information for selecting the suitable choice of elementary actions to compose a complex action. This contextual information taken from the environment via sensory system was used to estimate the future context resulting from the execution of the intended action, based on the information about the current state of the environment. Therefore, by viewing the whole perceptual process as a hierarchical chain of perceptual modules, we can say that the local organization of each stage of processing must obey the constraint that contextual properties should be present at each step. Or,

---

[2] Understand here mapping as topological organization.

[3] A holonomic constraint forces the propagation of geometric conditions among connected spaces; a non-holonomic one gives freedom for the connected spaces to follow only their own local conditioning.

that the transformations that produce invariant representations must bear the contextual properties relevant to produce the output required as input to the action system that lies at the other end.

As result of decades of research on learning processes for pattern recognition and vision, Tomaso Poggio and his collaborators have proposed a geometric theory for the perceptual processing, as recently reported in Poggio & Anselmi (2016). Their argument was based on their proposed principle of learning invariant representations. The idea can be roughly described as a way for reducing the complexity of learning, by imposing some constraints to the specification of the learning machines so that they perform their inferential tasks considering the invariance under specified transformations. Thus, at the first steps of visual perception, receptive fields formed by Hubel-Wiesel (HW) units would cope with invariance to translation, which is enough to handle local edges and small segments. Further steps that form wider receptive fields should cope with extra aspects of invariance, namely scale and rotation, besides translation, to handle lines and junctions. Finally, yet larger fields should handle shapes composed by the prior detected features, so the transformations should be class-invariant and learned through samples. There are many results and some predictions of their theory presented in that stunning work, nevertheless an important lesson that arises from their choice of transformations that led to invariances is exactly pointing to our proposed geometric model: the first steps of the perceptual process in their model employ just transformations that bring contextual properties, which are the affine ones. They are not related to the intrinsic conceptual nature of the object themselves, but to the geometrical properties that are invariants of the visual geometry. These transformations are mainly found in V1, which is a region shared by both the dorsal and the ventral streams. However, the class-specific invariances are expected to be found in further steps of the perceptual process in the ventral stream, and are just what is expected as a transformation whose invariants are related to conceptual aspects of the represented object, like a face, for instance, or several other examples provided by the reference. Our hypothesis is that a cognitive chain starts from this branch of the perceptual chain, the ventral stream, and progressively becomes more and more independent of contextual aspects and produces representations of an ontological nature, reflecting conceptual properties and relations of the visual world. Analogous reasoning can be applied to other forms of perception and to multimodal representations, which combine aspects originated from several perceptual channels. However, yet one important question remains: as the representations refer to more intrinsic conceptual properties of the objects and their relations and less or none to the contextual properties that have a physical basis, can one still talk about geometry of invariants in this case? How to envision such geometries?

Starting with a view inspired by the Erlangen Program and the work of Poggio and collaborators (Poggio & Anselmi, 2016), we suggest that cognition constitutes a space of processes that generates classes of invariants under class-specific transformations that operate on contexts internally generated, comprising possible relations involving the composition of prior detected invariants. More particularly, this space forms a category[4] made of processes corresponding to these classes of invariants and functors relating these processes. That such a mathematical structure can be viewed as a geometry in the sense of Klein ([1872] 1892) is argued by Marquis (2009). Anyway, category theory has been proposed as a potentially powerful model for cognition (Ehresmann & Vanbremeersch, 2007). The compositional properties of categorical operations, like products and coproducts, and the production of derived new structures like limits and colimits, have potential to furnish whole worlds built from small ones internally produced, and this process follows recursively, starting from rough representations of the external world that arise since the first moments of the experience of the agent. The question of how this representational internal world can make sense considering the richness and complexity of the actual external world is a very complex one and must be investigated in the line of natural transformations, which we do not consider here. Nevertheless, we could at least provide some preparatory topics concerned with the logical aspects involved in the satisfaction of basic constraints arising from the qualitative spatial structure of the world. These will be the first basic guides for establishing the foundational aspects of invariance, expected to be found in the construction of knowledge derived in the proposed fashion.

5. Qualitative Spatial Reasoning

Qualitative Spatial Reasoning (QSR) is a sub-field of Knowledge Representation in Artificial Intelligence (AI) that aims at the development of (formal) systems for representing and reasoning about spatial entities. The origins of QSR could be traced to the beginning of the 20th century with Alfred Whitehead's proposal for the development of a phenomenological theory capable of formally describing the world as picked out by the human perceptual system (Whitehead, 1929). In this phenomenological theory, "actual entities" are the primitive elements in the construction of a point-free geometry that makes explicit the basic relations (or principles) between these entities, allowing the derivation (or inferences) of the possible configurations of actual entities, given the underlying principles. Although there is much philosophical debate about what such theories should consist of, examples of point-free mathematical formalisms for the representation

[4] In the sense of mathematical category theory.

of space have been applied to provide the foundations of physics (Clarke, 1981), and as a tool in the development of artificial autonomous intelligent systems (Cohn & Renz, 2008).

In order to provide an example of a QSR formalism, we briefly introduce here the best-known formalism in AI, the Region Connection Calculus (RCC) (Randell, Cui & Cohn, 1992). RCC is a first-order theory with spatial regions (as "actual entities") and defined upon the primitive relation of connection between two such regions, denoted by C, which is a symmetric and reflexive binary relation. From this relation a set of mereotopological relations can be defined as shown in Fig. 10.1.

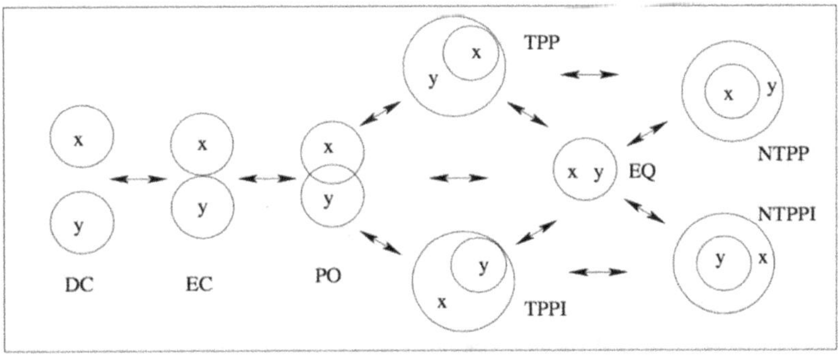

Figure 10.1: Relations in RCC8 (Randell, Cui & Cohn, 1992; permission for publishing granted by the authors).

Fig. 10.1 shows a conceptual neighborhood diagram for the RCC relations. A conceptual neighborhood diagram (CND) is a fundamental tool in QSR that expresses the smooth transitions of relations of a QSR formalism, i.e., between any pair of edge connected relations in the CND graph, there is no other relation that characterizes the state of the entities in the relation's arguments. Considering RCC, Fig. 10.1 expresses on the left-hand side two regions, x and y, that are disconnected: DC(x,y). From this state, the next possible relation (considering smooth transitions only) is externally connected: EC(x,y). After that, x and y can partially overlap: PO(x,y). From PO, three possibilities unfold: either x is equal to y, EQ(x,y), or x is a tangential proper part of y, TPP(x,y), or x is an inverse tangential proper part of y, TPPI(x,y). Finally, there are the relations NTPP(x,y) and NTPPI(x,y) which are similar to the previous ones, but this time there are no tangential regions. The set of relations constituted by {DC, EC, PO, EQ, TPP, NTPP, TPPI, NTPPI} is the jointly exhaustive and pairwise disjoint set known as RCC8.

Another important tool in QSR (at least for AI applications) is the transitivity table. Given three distinct domain individuals x, y and z, an entry in a transitivity table provides the transitive closure R3(x,z) (i.e., disjunction of the possible relations holding between x and z given) and a pair of dyadic spatial relations R1(x,y) and R2(y,z) of a spatial formalism (Randell, Cohn & Cui, 1992).

A set of qualitative spatial relations can also be understood as spatial constraints holding on domain objects. As presented by Ligozat (2011), consistency in a scenario described by a set of relations can be verified by enforcing algebraic closure (also known as Path Consistency) on a network of constraints. A constraint network is a pair (N,C), where N is the set of vertices (each vertex representing a domain element), and C is a set of constraints (i.e. spatial relations).

Let $C(i,j) \in C$ be a constraint on a vertex $(i,j) \in N \times N$. The network (N, C) is algebraically closed if for any triple $(i,j,k) \in N^3$ we have:

$$C(i,j) \subseteq \big(C(i,k) \circ C(k,j)\big).$$

Algebraic closure can be enforced by iterating

$$C(i,j) \leftarrow C(i,j) \cap (C(i,k) \circ C(k,j))$$

as long as a change occurs. As the set of basic relations is finite, this process becomes stationary after a finite time.

Similar ideas were developed for spatial formalisms based on direction, orientation, distance, size, and shape, among other aspects of space, as overviewed in Cohn & Renz (2008) and Ligozat (2011). Notwithstanding the claims in the current QSR literature that these formalisms are "closer" to the human characterization of space, recent experimental results comparing answers from human subjects and the formalism's inferences have proven otherwise (Rodrigues et al., 2016; Rodrigues et al., 2017). This discrepancy is not surprising, considering that the QSR formalisms were built upon abstract ideas about the physical space, leaving aside any aspect of the human (spatial) perception or cognition. The question of whether the mathematical tools developed in this field could be used to build a qualitative spatial representation based upon our spatial cognition is still open. The very nature of the human cognition is not yet completely understood, however some hints of it can be abstracted from work on cognitive psychology.

6. Spatial representation and advantaged cognition

Casati (2003) proposes the use of the results from cognitive psychology for supporting claims made by descriptive metaphysics (that aims at providing an exhaustive description of the world as perceived by the senses). This would provide the appropriate tools for building hypotheses about what people actually think. These hypotheses would then be based on the pieces of evidence (or at least on the constraints) underlying the basic mechanisms in cognition. Casati (2003) then cites a number of notions used in experimental psychological explanations (such as figure- ground articulation, salience, prototypicality, preference rule, preferential lexicalization, grounding, relevance) that are deemed to be prioritized (or, "advantaged") over other items in cognitive systems. In a sense, the advantaged object/feature plays the role of the "actual entities" cited in Whitehead's seminal work mentioned above (Whitehead, 1929). The task of the project described at the end of the previous section shares some similarities with this search for representational advantages in psychological research but, instead of providing the appropriate support for descriptive metaphysics, the goal is to collaborate with new foundations for the field of qualitative spatial reasoning. By assuming a formal system for artificial intelligent systems, this project escapes from the dilemma described in Casati (2003), which is the impossibility to make any claims about cognitive advantages, since these may be behavioral rather than contentual (i.e., we would be making assumptions about the content of a representation based on behavioral data). In an artificial system, content and behavior are transparent to the user.

Looking into cognitive psychology, Barbara Tversky (2011, p. 500) suggests that there is an interaction between the space we shape by living in and the shape of our own perceptual organization: "place and form constrain and convey meanings". One example of a visual feature becoming a feature of thought is the generalized tendency of organizing human space center-outwards (e.g., people sit by the center of a desk, the title of an article is centralized, and also in space analogies, such as "she is the center of attention"). This advantaged use of personal space may be due to the organization of the retina, as the point of greatest acuity in vision is located at its center. Gravity also imposes asymmetry in the vertical dimension: as it takes resources to overcome, a higher position is usually associated with health, wealth, strength, etc. (e.g., the place of a gold-medalist in a podium); whereas horizontally there is no privileged dimension (handedness is an exception that seems to be primarily behavioral rather than perceptual; cf. Tversky, 2011). Besides, the use of diagrammatic representations to convey (non-spatial) abstract relations (e.g. force diagrams in basic-level physics textbooks) suggests that spatial reasoning also forms a foundation for more abstract reasoning (Tversky, 2005). Experimental data from humans locating objects in space suggests that a mental spatial framework is constructed from extensions of the three axes of their body: head/feet, front/back, left/right. The responses to head/feet are the fastest, followed by front/back and then by left/right.

Tversky (2005, p. 216) suggests a catalogue of candidates for advantaged elementary properties of representations:

- static properties of entities (figure/ground, symmetry, shape, internal configuration, size, color, texture);
- relations between static entities (with respect to frame of reference: location, direction, distance; or, with respect to other entities);
- relations of dynamic and static entities;
- performing transformations on entities;
- performing transformations on self.

Besides finding the relevant candidates for advantaged object/property, there is the more complex task of assimilating in our (formal) theory the inconsistencies, approximations, vagueness and distortions in memory and judgement for shapes, configurations of objects, distances, and directions, that are accommodated by human cognition (as reviewed in Tversky, 2005). The key question that we pursue here has been expressed by Tversky (2005, p. 224): "what might a representation that captures all these distortions look like?"

The field of qualitative spatial reasoning (briefly described in section 5 above) seems to have cut into pieces the various possible candidates of advantaged elementary properties, whereby each QSR formalism is a mathematical specialization of one particular modality in this catalogue. One of the obvious undesired consequences of this specialization is that it becomes far too complex to merge formalisms in order to allow representation and reasoning with combinations of elementary entities (Gabelaia et al., 2005). The root of the problem seems to be that research in QSR have not taken into account the actual reasoners (as individuals) or the environment they live in, and thus have overlooked the constraints imposed by this pair, that cuts down "irrelevant" features of representation (and reasoning) about space.

In recent work, we have used the "qualitative ego-sphere" (Rodrigues et al., 2016), a formalism for representing qualitative location that discretizes the world around a visual agent into sectors (analogous to a spatial framework theory, as in Franklin & Tversky, 1990), as a basic language for communicating spatial expressions between humans and artificial agents. Experiments with human volunteers suggested that the human description of space into sectors had an average of 65% of agreement with the qualitative discretization provided by the "qualitative ego-sphere". The answers provided by the subjects were vague approximations of the object locations (as given by the formalism). One of the reasons for that was the rigidness of the formal system, whose parameters were set beforehand (without the preliminary analysis of how humans perceive space). Currently, we have been applying machine learning techniques to find a better model. However, the use of vague (or polysemic) concepts to locate objects cannot be tackled with a direct application of machine learning. In another work (Rodrigues et al., 2017), we introduced a number of qualitative spatial relations in order to provide precise meanings for a spatial preposition. The resulting formalism aims at grounding the meaning of a polysemic preposition considering the scenarios in which it is applied. Through the use of precisifications, and the typical uses of the language (obtained from the application of a questionnaire on human subjects), the research showed that a fragment of supervaluation semantics, initially employed in vague languages, was successful to model polysemy. But then, again, by applying the modeling of language to a model of cognition, we incurred in the sin of assuming contentual knowledge (i.e. what actually lies beneath cognition) with the representation of information (language).

Bibliography

Casati, R. (2003). Representational advantages. *Proceedings of the Aristotelian Society,* 103(1), 281-298.

Clarke, B.L. (1981). A calculus of individuals based on 'connection'. *Notre Dame Journal of Formal Logic,* 22(3), 204-218.

Cohn, A.G. & Renz, J. (2008). Qualitative spatial representation and reasoning. In: van Hermelen, F.; Lifschitz, V. & Porter, B. (eds.). *Handbook of knowledge representation.* Amsterdam: Elsevier, pp. 551-596.

Cooper, R.P. (2016). Schema theory and neuropsychology. In: Arbib, M.A. & Bonaiuto, J.J. (eds.). *From neuron to cognition via computational neuroscience.* Cambridge (MA): MIT Press.

De Waal, F. (2016). *Are we smart enough to know how smart animals are?* London: Granta.

Ehresmann, A.C. & Vanbremeersch, J.P. (2007). *Memory evolutive systems: hierarchy, emergence, cognition.* Amsterdam: Elsevier.

Franklin, N. & Tversky, B. (1990). Searching imagined environments. *Journal of Experimental Psychology,* 119(1), 63-76.

Gabelaia, D.; Kontchakov, R.; Kurucz, A.; Wolter, F. & Zakharyaschev, M. (2005). Combining spatial and temporal logics: expressiveness vs. complexity. *Journal of Artificial Intelligence Research,* 23(1), 167-243.

Goodale, M.A. & Milner, A.D. (1992). Separate visual pathways for perception and action. *Trends in Neurosciences,* 15(1), 20-25.

Hubel, D.H. & Wiesel, T.N. (2014). *Brain and visual perception: the story of a 25-year collaboration.* Oxford: Oxford University Press.

Jeannerod, M. (2006). *Motor cognition.* Oxford: Oxford University Press.

Johnson, M.H.; Mareschal, D. & Csibra, D. (2008). The development and integration of the dorsal and ventral visual pathways in object processing. In: Nelson, C.A. & Luciana, M. (eds.). *Handbook of developmental neuroscience.* 2nd ed. Cambridge (MA): MIT Press, pp. 467-498.

Klein, F. (1892). A comparative review of recent researches in geometry. Transl. M.W. Haskell. *Bulletin of the New York Mathematical Society,* 2, 215-24. Available at: https://arxiv.org/pdf/0807.3161.pdf. German original: 1872.

Ligozat, G. (2011). *Qualitative spatial and temporal reasoning.* New York: Wiley.

Marquis, J.P. (2009). *From a geometrical point of view.* New York: Springer.

Mishkin, M. & Ungerleider, L.G. (1982). Contribution of striate inputs to the visuospatial functions of parieto-preoccipital cortex in monkeys. *Behavioral Brain Research,* 6(1), 57-77.

Poggio, T. & Anselmi, F. (2016). *Visual cortex and deep networks: learning invariant representations.* Cambridge (MA): MIT Press.

Randell, D.; Cui, Z. & Cohn, A. (1992). A spatial logic based on regions and connection. In: Nebel, B.; Swartout, W. & Rich, C. (eds.). *Proceedings of the 3rd International Conference on Knowledge Representation and Reasoning.* San Mateo (CA): Morgan Kaufmann, pp. 165-176.

Randell, D.A.; Cohn, A.G. & Cui, Z. (1992). Computing transitivity tables: a challenge for automated theorem provers. In: Kapur, D. (ed.). *Automated deduction – CADE 11: 11th International Conference on Automated Deduction.* Lecture Notes in Computer Science, 607. New York: Springer, pp. 786-790.

Rodrigues, F.M.E.; Santos, P.E. & Lopes, M. (2016). Communication of spatial expressions on multi-agent systems using the qualitative ego-sphere. In: *12th IEEE International Conference on Control and Automation (ICCA 2016)*, pp. 25-30.

Rodrigues, E.; Santos, P.E. & Lopes, M. (2017). Pinning down polysemy: a formalization for a Brazilian Portuguese preposition. *Journal of Cognitive Systems Research*, 41, 84-92.

Tomasello, M. & Call, J. (1997). *Primate cognition*. Oxford: Oxford University Press.

Tversky, B. (2005). Visuospatial reasoning. In: Holyoak, K. & and Morrison, R. (eds.). *The Cambridge handbook of thinking and reasoning*. Cambridge (GB): Cambridge University Press, pp. 209-241.

——— (2011). Visualizing thought. *Topics in Cognitive Science*, 3, 499-535.

Vernon, D. (2014). *Artificial cognitive systems*. Cambridge (MA): MIT Press.

Whitehead, A.N. (1929). *Process and reality*. New York: Macmillan.

Chapter 12

How complex behavior emerges from spikes

João Ranhel
Departament of Electronics and Systems
Federal University of Pernambuco (UFPE), Recife, PE, Brazil
jranhel@upgames.com.br

Emilio Del Moral Hernandez
Polytechnic School
University of São Paulo (USP), São Paulo, SP, Brazil
emilio@lsi.usp.br

Marcio Lobo Netto
Polytechnic School
University of São Paulo (USP), São Paulo, SP, Brazil
marcio.netto@usp.br

Abstract

Neural cell assemblies emerge on spiking neural networks with propagation delay (SNNwD) because spikes coincide on specific neurons, causing them to fire collectively. Interactions among neural assemblies may perform logical functions, bistable loops, finite-state automata (FSA) and other information processing functions, called here 'functional blocks'. In this chapter, we describe how to design functional blocks with spiking neurons, and how to create algorithms by joining FSA, memory, decision-making circuits, and other blocks. Algorithms generate stereotyped responses that may lead to intelligent and complex cognitive behaviors. In order to design algorithms, we used neural engineering and a framework called neural assembly computing (NAC), an approach concerned with how neural assemblies process information. We review NAC fundamentals and introduce new concepts, such as gating, sampling, analog memory, FSA, and decision-making blocks. We also implement an algorithm in order to demonstrate the concept. In NAC designs, an important guideline is that all the system components must be embedded within the network, therefore no external artifact can operate over the SNNwD. Spiking neurons, synaptic weights, and delays are all the elements with which designers can count on to implement functions, mechanisms, and rules in NAC. Our research pointed in the direction of implementing plasticity mechanisms in NAC, as well as implementing an astrocytes-neuron network as a mechanism for co-opting neurons in order to create or modify existing cell assemblies.

1. Introduction

Neural engineering is a field that uses engineering techniques to exploit properties of natural neural systems, working at the interface of biological neural systems and artificial neural networks (ANN) (Eliasmith & Anderson, 2004). Certain approaches in neural engineering try to reproduce properties and phenomena found in live neural tissues. Animals have innate neural systems capable of performing astonishing *information processing*, in order to represent the world, classify things (food, predators, etc.), or deal with relations (distances, timing, sizes, etc.). While many animals have modest neural systems with minor learning abilities (Smith, 2000; Kandel et al., 2000; Purves et al., 2004), others have high-level cognitive capabilities (Bekoff et al., 2002). It is common to hear that animals have *hardwired* circuits that trigger actions such as to escape, forage, attack, etc., which we call *stereotyped algorithms*. How such algorithms are executed in neural networks is under investigation. We have designed artificial neural circuits that execute predefined responses mimicking animal algorithms. Undoubtedly a great value is assigned to understanding how such stereotyped responses may be reproduced, as well as how cognitive behavior can emerge from basal hardwired structures. This knowledge may be used in robotics, information processing, and in artificial intelligence.

Computer scientists and engineers have obtained valuable knowledge about information processing. Logical functions, memory, decision-making, finite-state automata (FSA), among others, are well-known functions in the field of information processing, being used for creating computing systems. We consider that nervous systems rely on hardwired circuits, hierarchically controlled, in the same way engineers rely on hierarchically interconnected logical blocks to form complex information processing systems, even considering that they may differ in their basis. Hence, we have coupled basic functional neural blocks to create more complex neural computation.

In this paper, we show how to combine neural blocks in complex structures that perform stereotyped behaviors. Our aim is to embed within the ANN all necessary mechanisms to implement these abilities with no external computational artifact interfering on the ANN. We have used *spiking neural networks with propagation delay* (SNNwD), which intends to be closer to the nervous systems than previous generations of ANN (Paugam-Moisy & Bohte, 2010). We do not claim that biological systems work the way we describe herein, although our approach is strongly based on biological inputs. There is much to be learned from biological systems. As an example, the way neurons codify the world is a top concern in neuroscience (Cessac et al., 2010; Kumar et al., 2010; VanRullen et al., 2005). Here, we opted to deal with two neural coding schemes: fire rate and population coding.

2. Overview of Neural Assembly Computing

Neural Assembly Computing (NAC) is an approach that aims to explain how sets of cell assemblies represent and process information, how they perform algorithms, and how intelligent behaviors emerge from them (Ranhel, 2012). A key concept in NAC is that neurons represent internal and external signals, objects, signs, states, etc., by means of cell assemblies. Neural assemblies are sets of neurons that, somehow, fire together. Interactions among neural assemblies perform logical functions. Assemblies can also form loops that memorize information, as well as function as pacemakers.

The notion that groups of neurons (instead of single cells) represent, memorize, and underlie brain processes dates back a long time. For instance, in 1949, Donald Hebb proposed that concepts are represented in nervous systems by the activity of partially overlapping neural groups, and the co-activation of cell assemblies are responsible for representing things (Hebb, 2002).

2.1 Triggering cell assemblies

In the early 20th century, Charles Sherrington argued that groups of neurons would be responsible for the activation of the muscles, which was later supported by evidence (König et al., 1996). People long thought of neurons as calculators of averages, however, in 1982 Abeles proposed an alternative concept (Abeles, 1982; König et al., 1996): neurons may act as 'coincidence detectors'.

The spike propagation delays create conditions for the emergence of neural cell assemblies. Spikes do not spread instantaneously along the axons. The neuroscience literature reports action potential velocities in large ranges, such as 1 to 100 m/s (Kandel et al., 2000), 150 m/s (Purves et al., 2004), or 0.15 m/s (Izhikevich et al., 2004), which may cause delays of approximately 10 ms even in short neural ensembles.

Considering neurons as coincidence detectors (Abeles, 1982), we may simplify the synaptic weight calculation by determining a post-synaptic *perturbation* (Θ) that guarantees a single spike in a neuron, a method also used by Wickens & Miller (1997). 'Perturbation' here is synonymous to excitatory or inhibitory post-synaptic potential (EPSP/IPSP) (Kandel et al., 2000; Purves et al., 2004; Arbib, 2003). In other words, the sum of coincident spikes from all fired pre-synaptic neurons causes Θ in a post-synaptic neuron at a certain instant. Let us consider a neuron (a member) in a post-synaptic assembly *B* that receives *n* pre-synaptic spikes from neurons in an assembly *A*. The synaptic weight can be calculated by

$$w_{ij} = \Theta / n \ , \qquad (12.1)$$

where w_{ij} is the synaptic weight from *i* (pre-) to *j* (post-synaptic neuron), *n* is the number of members in *A*, and Θ is the EPSP necessary for firing the neuron member in *B*.

In order to consider a set of neurons *B* responding as a cell assembly, all neurons in *B* are connected to all neurons in *A*. We say *A* is fully connected to *B* when all neurons from an assembly *A* are connected to all neurons in an assembly *B*. It does not mean that neurons from *A* are only connected to *B*, nor that neurons in *B* only receive connections from *A*. Instead of thinking of neural assembly computing (NAC) in terms of topologies, think of NAC as *events* happening in a SNNwD, although many of such events are favored by appropriated topologies.

2.2 Logical functions and memory

The combination of propagation delays, synaptic weights, and the network connections makes spikes spread out and reach many neurons at different instants. Consequently, sets of neurons (the assemblies) are reached by simultaneous stimuli and fire together, in synchronicity or in a time-locked pattern. Fig. 9.1 shows a raster plot with neural assemblies denoted K_0, K_1, K_2, and so on, each assembly with 20 neurons.

Sometimes spikes from an assembly *A* or from an assembly *B* can independently trigger a third assembly *C*. This means that *A* OR *B* can trigger *C*, which uses the logical function OR (denoted $C \leftarrow A \vee B$), a *disjunction*. This is illustrated by K_{10} in Fig. 12.1, which is triggered by $K_2 \vee K_7 \vee K_8$.

On the other hand, due to weak synaptic connections A→C, spikes from *A* may not be strong enough for causing Θ in *C* neurons. Equally, due to weak synaptic connections *B*→*C*, spikes from *B* may not trigger *C* neurons. However, when spikes from *A* AND *B* coincide they cause Θ and trigger all neurons in *C*, which performs an AND logical function (denoted $C \leftarrow A \wedge B$), a *conjunction*, illustrated in Fig. 12.1 by K_{12}, which is triggered by $K_4 \wedge K_8$.

A reader used to Boolean algebra may be confused with our notation, because, for an AND logical function, inputs *A* AND *B* must remain actively *high* (logic level '1') to obtain the output *C high*. But, spiking neurons fire for a brief instant (≈1 ms). Thus, assemblies execute logical functions (AND, OR, etc.) only during the ephemeral time in which the assembly occurs. Therefore, we chose to use *expressions* instead of *equations* in this article, because equation would mean that the formula is valid all the time, but in NAC formulas are valid only at the time a specific event occurs.

After a *cell assembly event*, the resulting spikes become *travelers*. In SNNwD, information travels along axons most of the time, resulting in what we called *branching*: an assembly can trigger more than one assembly, creating ramifications of the original information.

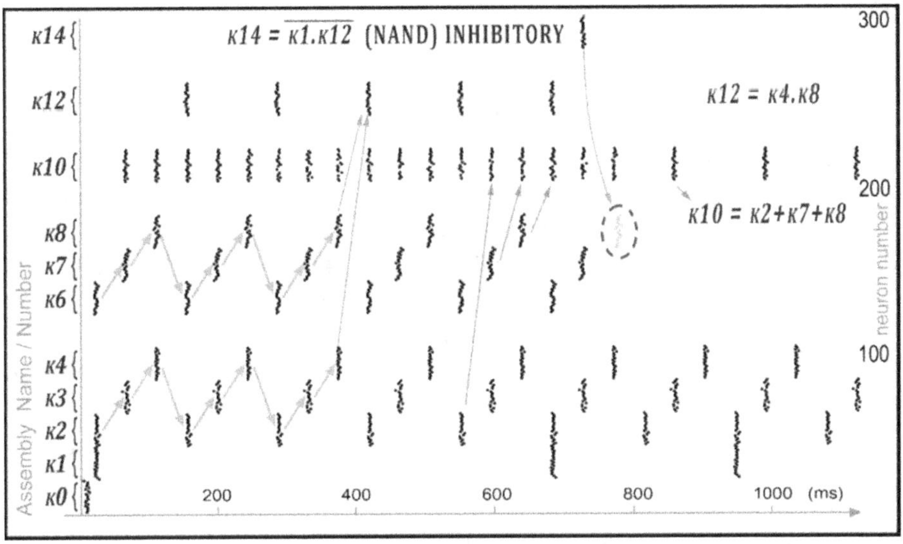

Figure 12.1: Basic functions in neural assembly computing. Assemblies are groups with 20 neurons. Assemblies K_2 - K_3 - K_4 perform a bistable loop, as well as K_6 - K_7 - K_8. Assembly K_{10} performs the OR function ($K_{10} \leftarrow K_2 \vee K_7 \vee K_8$), also written with "+", whereas K_{12} performs an AND function ($K_{12} \leftarrow K_4 \wedge K_8$), also written with "·". Assembly K_{14} performs a NAND function ($K_{14} \leftarrow \overline{K_1 \wedge K_{12}}$) by inhibiting K_8. (All figures made by the authors for this publication.)

Now consider that an event (a smell, a vision, a pain, an internal sign such as hunger, etc.) triggers an assembly. This means that such assembly *represents* the event that causes it. Considering that assemblies stand for *something*, the nervous systems should not let such events simply disappear, because they might carry important information. We may expect Nature to have found a way to maintain information, for instance, by *reverberation* among assemblies: an assembly *A* triggers an assembly *B* that triggers back *A*, forming a loop, such as $K_4 \leftarrow K_3 \leftarrow K_2 \leftarrow K_4$ and $K_8 \leftarrow K_7 \leftarrow K_6 \leftarrow K_8$ in Fig. 12.1 (see Ranhel, 2012, for more complex loops).

We call these reverberating loops *bistable neural assemblies* (BNA) because they can be of two states: ON (active) or OFF (inactive). As soon as we trigger a BNA it can retain one 'bit' without demanding any plasticity mechanism. This is a kind of memory sketch, or an instantaneous short-term memory. Once a BNA is a loop it would remain active indefinitely. In this case, another type of assembly playing the *veto* or *inhibition* role is necessary. An assembly can *dismantle* a BNA or any branch by inhibiting a participant assembly. In this case, inhibitory assemblies execute a function equivalent to the NOT logical function.

There can be cases when two assemblies may be necessary (AND) in order to inhibit a third one, and this conjunction executes a NAND-like logical function. The same is valid for an OR associated to inhibition, which creates a NOR-like function.

3. Finite-state automata

A finite-state automaton FSA (or finite-state machine) is a mathematical abstraction, a computing machine with finite memory, which is defined in terms of a finite set of states, an initial state, a set of input symbols (alphabet), and a transition function set that maps input symbols and current states to the next machine states (Hopcroft et al., 2007; Berwick et al., 2011).

FSAs can only be in one of a finite number of *states* at a time. A specific condition/event may cause a *transition* from one state to another. Therefore, a FSA is defined in terms of a set of states and a set of triggering conditions.

Finite-state automaton is a well-established concept in computer science. FSA are often separated in classes of formal languages they can recognize, for instance, finite, regular, context-free and context-sensitive languages (see Hopcroft et al., 2007; Berwick et al., 2011).

3.1 FSA and biological systems

Central pattern generators (CPG) are neural kernels capable of generating patterned rhythms, for example, CPGs generate patterned movements, such as the gaits for legged locomotion in animals (Ijspeert, 2008). It is possible to relate CPG and FSA: each CPG firing pattern may be considered a machine state; and inversely, each automaton state may generate an output pattern that commands something, in the same way as CPGs do.

Some researchers investigate FSA as pattern recognizers in biological stereotyped behaviors, such as speech recognition and speech production in humans and animals; for example, the syntax of songbirds was studied using FSA as a basal component (review in Berwick et al., 2011).

On another research front, Yuste and co-workers discovered similarities among neocortical circuits, vertebrate spinal cord and brainstem CPGs. Similarities were found in many levels of analysis: biophysical, pathological, dynamical, anatomical, and developmental (Yuste et al., 2005). Based on such findings, the authors argue that the neocortex can be made of special types of CPGs, which are Hebbian cell assemblies highly specialized in learning, storing or retrieving memories. They argue that central nervous system evolution represents the progressive encephalization of motor rhythms. The widespread occurrence of plasticity mechanisms in the neocortex, associated to different neuron types, may lead to inferences that in the cortex preponderate ever changing circuits; thus, the neocortex resembles a liquid machine. However, Hebbian plasticity does not preclude the existence of predefined classes of neurons with hardwired topology and specific dynamics. They conclude that such circuitry may have predetermined structures with relatively little plasticity (Yuste et al., 2005). These ideas are important neuroscientific support to our framework. This is exactly how we design CPGs and finite-state machines on NAC. Until now, our networks resemble hardwired circuits that perform finite-state automata with little plasticity, but that interact with each other.

3.2 Implementing a FSA on Neural Assembly Computing

Let us create a finite-state automaton (FSA) that deals with a two-symbol alphabet (0 and 1), which can recognize a pattern sequence, let us say "100". Every time this pattern occurs in an input stream, the automaton must show the occurrence. The first step is to draft a state diagram, as shown in Fig. 12.2a. In this diagram, we represent states by circles, each one labeled with a name. S_0 is the initial state. A bistable neural assembly (BNA) K_2 to K_6 creates the rhythm for the network, mimicking central pattern generators. BNAs are also responsible for memorizing each automaton state. A detailed description of the raster plot in Fig. 12.2b was published in Ranhel (2013).

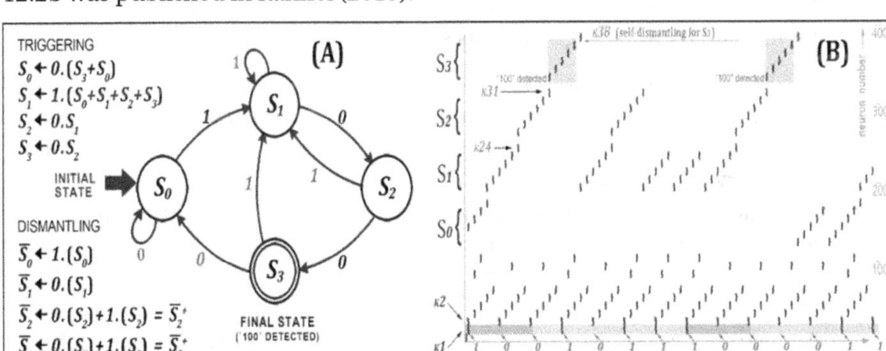

Figure 12.2: (a) State diagram and formulas for a FSA that detects an input pattern "100". The top formulas describe how to trigger new states whereas the formula below describes how to dismantle the current state (when necessary). (b) Raster plot of a FSA detecting the pattern "100" from an input string. Each assembly has 10 neurons. At each K_2 occurrence, one input bit is presented to the FSA.

While the automaton is in S_0, if a '0' comes as an input, the FSA remains in S_0, otherwise, an input '1' causes a transition to the S_1 state. Insofar as the machine is in S_1 and an input '0' comes, a transition puts machine in S_2. In this case, it has detected "10". Once the automaton is in S_2 and another '0' comes, the machine goes to the final state S_3. As the FSA achieves S_3 it may stop or it may continue indefinitely detecting the same pattern, which is the case here. Thus, once in S_3 and an input '0' comes, the FSA restarts at S_0; otherwise, an input '1' has been detected, and the cycle restarts from S_1.

From S_0 or S_1 to S_3, the machine must follow a forward flow in order to detect "100". However, designers must provide all other possibilities when designing a FSA. Thus, once in S_0 and an input '0' comes, the machine must remain in S_0.

If the automaton is in S_1 and an input '1' comes, the machine stays in S_1. Once in S_2 ("10" detected), if a '1' comes, the sequence is broken ("101"), so from S_2 the automaton must go to S_1.

After defining the state diagram, we obtain the *expressions* for all machine transitions (listed in Fig. 12.2a). For instance,

$$S_0 \leftarrow (0 \wedge S_0) \vee (0 \wedge S_3) \tag{12.2}$$

can be read: S_0 is triggered whenever the FSA is in S_0 and '0' comes as an input, OR whenever it is in S_3 and a '0' comes. We also obtain expressions for inhibitions; e.g. for S_0 it is

$$\bar{S}_0 \leftarrow 1 \wedge S_0 \,, \tag{12.3}$$

which is read: S_0 is dismantled (NOT S_0) whenever the machine is in S_0 state and an '1' comes as an input. Once in S_0, as the FSA receives a '1' the S_0 state must dismantle itself, because it is leaving this state to change to S_1.

As discussed earlier, assemblies represent objects, values, bits, and states. In neural assembly computing, finite automata memorize states into BNAs. In this sense, "100" may signify a sequence of events with a semantic meaning: "if an animal is hungry '1', no food is perceived '0', and no obstacle is present '0', then it triggers another FSA in the network, e.g., for foraging". The automaton above would do that as it reaches the S_3 state.

4. Algorithms in spiking neural assemblies

The term algorithm is generally used for defining a set of finite steps executed by a machine to perform a task. They frequently require decisions and loop structures that control the transitions. As examples, the IF–THEN–ELSE is a common decision-making structure used in computer programs. In the same way, the DO–WHILE is a common loop structure. Before explaining how to create algorithms, let us examine how to implement some structural blocks in neural assembly computing (NAC).

4.1 Sampling, decisions and loops

Some structures used to construct finite-state automaton (FSA) in NAC are intrinsically loops (e.g. $K_6 \leftarrow K_5 \leftarrow K_4 \leftarrow K_3 \leftarrow K_2 \leftarrow K_6$ in Fig. 12.2b). Loops are repeated until an assembly member is inhibited. From another point of view, the sequences: $\{S_0, 1, S_1, 0, S_2, 0, S_3, 0, S_0\}$, and $\{S_0, 1, S_1, 0, S_2, 0, S_3, 1, S_1\}$ are both loops that may occur on the FSA described in Fig. 12.2. Pacemakers are also loop examples; thus, loops are not a problem for NAC designs.

Two decision-making blocks, described later in this text, may be easily instantiated in NAC, but they require operations that forego the decision process: to get samples (*sampling*) and to store (*memorizing*) information. Let us analyze how we create two auxiliary neural circuits for decision-making: one is a topology that functions as a gate, or a sampler for analog inputs; the other is the analog memory structure, responsible for retaining analog values in the population code scheme.

4.2 Analog gates

Fig. 12.3 shows a topology for an analog gate block. Consider an assembly G with k neurons. Each g_j neuron in this 'gate' is connected to a single input i_j neuron that belongs to a pre-synaptic assembly F. The synapse weight from each input neuron to each G is calculated by $w_j = \Theta/2$.

Consider a *pacemaker* P, P_A, and P_B (a bistable neural assembly), all neurons from P connected to each g_j neuron. Considering eq. (12.1), $w = \Theta/n$, for this gate we use $w_{ij} = 1/2(\Theta/n)$; all neurons i from P are connected to each neuron j on G. This means that when all P neurons fire, each neuron in G receives an excitatory post-synaptic potential (EPSP) of $\Theta/2$, which is not enough for triggering any g_j neuron. However, when spikes from P coincide with a spike from the f_j neuron at the g_j neuron, this results in a perturbation Θ in this g_j neuron, calculated as

$$\Theta_k \leftarrow \Theta/2_{(p)} + \Theta/2_{(ij)} \ . \qquad (12.4)$$

Therefore, spikes from inputs can only propagate through the paired f_j, g_j neurons when P opens the gate. In other words, P causes a bias in all g_j neurons, enabling the gate G, and each analog f_j neuron triggers its paired g_j neuron.

This topology takes a *sample* of all spiking members in *F* controlled by the event *P*. We call this topology *analog gates* (AG) because, in population code, the number of active members when *P* occurs represents an analog information.

Fig. 12.4c shows, at its very bottom plots, examples of two analog inputs AI_1 and AI_2, emulating external stimuli. The simulator randomly generates the inputs and two sine functions modulate which members are active. Neurons from these populations are connected to AGs that sample their inputs, generating *X* and *Y*. Samples are synchronous with A_3, but due to propagation delays, they are seen when A_4 is firing.

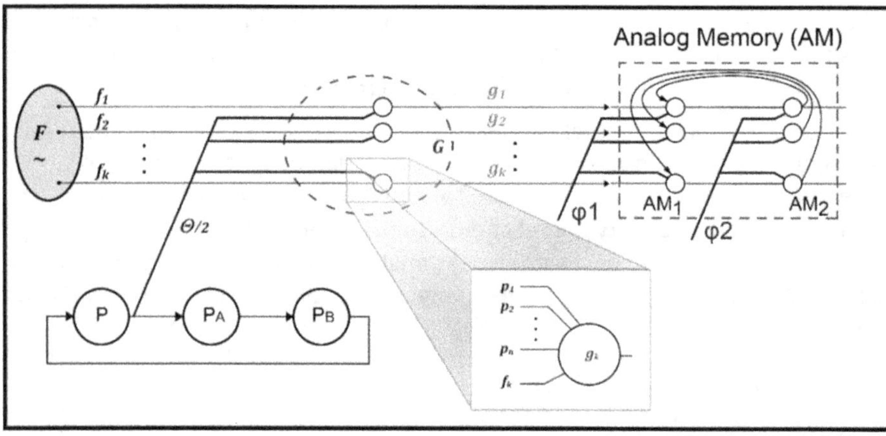

Figure 12.3: Analog gate and analog memory: all neurons in assembly *P* are connected to all neurons in gate *G* by a synaptic weight Θ/2. Each neuron from input *F* is connected to a single neuron in G also by Θ/2. As *P* fires, only a firing neuron f_k can trigger its pair g_k neuron. The gate is associated to a bistable loop that realizes an analog memory. The φ_1 and φ_2 synchronize the memory reverberation.

Biological *oscillation-gating* functions (see Buzsáki & Draguhn, 2004) inspires this biased-gate topology. It is outside of the scope of this article to go deeper into the synchronization of external stimuli. Synchronism in inputs is a large issue itself (see Brette, 2012).

4.3 Analog memory

The *analog memory* (AM) concept is an association of two network arrangements: analog gates and bistable neural assembly loops. The function of AMs is to memorize analog values codified in the population code scheme.

In Fig. 12.3, consider all g_k neurons as outputs of the gate that must be memorized. Assemblies are ephemeral events, thus, the network must keep the spikes reverberating within a loop. Hence, suppose all g_k output neurons are copied in the assembly AM_1 at the time φ_1 (Fig. 12.3). Similar to the gate topology, the signal φ_1 provides an EPSP equal to $\Theta/2$. Each neuron g_k is also connected by w_{if}=$\Theta/2$ to a paired neuron in AM_1, hence, only the neurons that fired in the analog gate (the firing g_k neurons) will fire the correspondent neurons in AM_1. Moreover, all neurons from AM_1 are connected also by w_{if}=$\Theta/2$ to the neurons in AM_2, which fire synchronized to φ_2. The signal φ_2 also provide an EPSP equal to $\Theta/2$. Therefore, only those neurons that fired in AM_1 will fire in AM_2. The reverberation loop is closed because all neurons in AM_2 are connected back to the neurons in AM_1 by the same synaptic weight. As long as φ_1 and φ_2 remain reverberating the loop latches the spikes.

In summary, *P* opens the gate for external firing neurons. In the input population *F*, only those neurons that are firing force their paired neuron g_k in the gate to fire. Then, these spikes propagate to the assembly AM_1 (that fires synchronous to φ_1), and reach the assembly AM_2 synchronized to φ_2. Spikes from AM_2 then go back to AM_1 and the loop remains reverberating if φ_1 and φ_2 remain active. We call this an 'analog memory' because it maintains a set of neurons firing that is proportional to the size of the input population *F* when the gate is opened. It is possible to erase the memory contents by inhibiting φ_1 or φ_2, and it is possible to change the content by opening the gate again, as described above.

5. Decision-making, threshold and hysteresis

Consider an analog assembly *X* with η neuron members ($1 \leq \eta < 2k$), where *k* is the number of members in digital assemblies (DA), such as in the *pacemaker* created by A_3, A_4, and A_5, as shown in Fig. 12.4a. Consider all neurons from *X* and A_3 connected to an assembly *R* by a synaptic weight $w/3$, with w calculated as in eq. (12.1). *R* has *k* neurons and behaves as a digital assembly; hence, *R* can fire whenever A_3 is firing, but only after a certain number of neurons in *X* are firing, calculated by

$$\Theta_R^t \leftarrow \left(\eta_X^t . w + k_{A3}^t . w\right) / 3 \ , \qquad (12.5)$$

where Θ_R^t denotes the total perturbation in any *R* neuron at time *t*, the term $\eta_X^t . w$ denotes the perturbation caused by η neurons firing in *X* at time *t*, and $k_{A3}^t . w$ denotes the portion of perturbation caused by *k* neurons firing in A_3 at time *t*. This topology guarantees that *R* is triggered only after $\eta \geq 2/3$ of *X* members are firing. Let us consider this the *higher* threshold.

Now, let us introduce hysteresis: R is connected to a DA named A_{11}. Then, A_{11} is connected to another DA named A_{12} for timing compatibility, and then A_{12} is connected back to R by synaptic weights that also cause $\Theta/3$ in R. This means that, via A_{11} and A_{12}, R causes EPSP in its own neurons, increasing Θ_R^{t+3} by $\Theta/3$. Note that it takes Δt for spikes from R to reach A_{11}, it takes Δt for A_{11} spikes to reach A_{12}, and Δt for A_{12} spikes to reach R again; therefore, the next event in R is delayed by $3\Delta t$. Such positive feedback results in that, once R is ON, the network can only dismantle R when X has less than 1/3 of its members active. Let us consider this the *lower* threshold. This is the condition for R to turn OFF. Then, Θ_R after $3\Delta t$ is

$$\Theta_R^{t+3} \leftarrow \left(\eta_X^{t+3}.w + k_{A3}^{t+3}.w + k_{A12}^{t+3}.w\right)/3 \ , \qquad (12.6)$$

where $k_{A12}^{t+3}.w$ denotes the EPSP that k neurons firing in A_{12} cause in R at the time $t^{+3\Delta t}$; but only when the assembly R is ON, otherwise this term is zero, as in eq. (12.5).

In Fig. 12.4c, the *threshold and hysteresis* (*T&H*) network is observed to act over the X samples (see R, A_{11}, and A_{12}). Note that R only turns ON after reaching the *higher* threshold, and it keeps this decision until the input reaches the *lower* threshold.

5.1 The greater of two

Let us consider two assemblies X and Y, outcomes of two AGs, both with η neurons ($1 \le \eta < 2k$), where k is the number of neurons in A_3, A_4, and A_5, which are DAs used as a pacemaker. Then consider X and Y connected to the A_8 and A_9 assemblies, as in Fig. 12.4b, both A_8 and A_9 generate digital outputs. X and Y are synchronous to A_3. In this case, it is possible to determine the perturbation in A_8 and A_9 because they are connected by reciprocal excitatory $+w$ and inhibitory $-w$ connections. We calculate Θ for them by

$$\Theta_9 \leftarrow (k_X - k_Y).w \quad , \qquad (12.7)$$

$$\Theta_8 \leftarrow (k_Y - k_X).w \quad . \qquad (12.8)$$

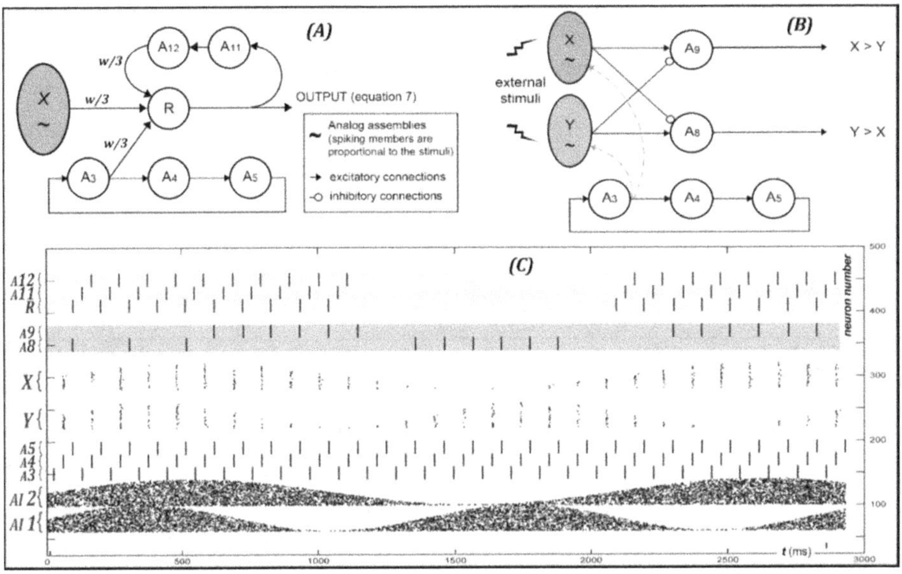

Figure 12.4: (a) Decision-making network topology with double threshold. (b) Topology for decision-making 'greater of two' network. (c) Raster plot showing these two circuits working in neural assembly computing. Each assembly has 20 neurons. For details, refer to the text.

This means that perturbations in A_9 neurons depend on the difference between the number of excitatory firing members in X and the number of inhibitory firing members in Y. The opposite is valid for A_8.

Consider multiplying w by a constant α ($2 \leq \alpha \leq 10$), which allows controlling how large the difference between X and Y must be. A small difference in the number of firing members multiplied by a great α may cause the necessary Θ for firing A_9 or A_8. Experimentally, we found $\alpha=4$ to be a good value.

Bearing in mind that A_9 and A_8 are digital assemblies that fire in an all-or-none fashion, A_9 fires when X is slightly greater than Y, whereas A_8 fires when Y is slightly greater than X. This neural arrangement can implement the *greater of two* (GoT) function. For analog samples codified as a population code, the GoT function can determine which is the greater.

In Fig. 12.4c, the results (A_8 and A_9) of two samples X and Y are applied to the network of Fig. 12.4b. We observe that when the analog inputs are similar (e.g. at $0<t<600$ ms), the outputs A_8 and A_9 may be silent or one may sporadically fire. When the inputs are quite different, then the response is very clear: A_9 firing when $X>Y$ or A_8 firing for $Y>X$.

5.2 Constructing an algorithm

Logical functions associated to decision blocks and finite-state automata (FSA) allow us to create algorithms. To exemplify how neural assembly computing may underlie an algorithm, let us consider an agent searching for something. Consider the agent has an internal information of how much it needs, let us say, *water*. Above a certain limit, this need drives the agent to search for water, directing its behavior for achieving this goal. Suppose an analog assembly X represents the intensity of the need.

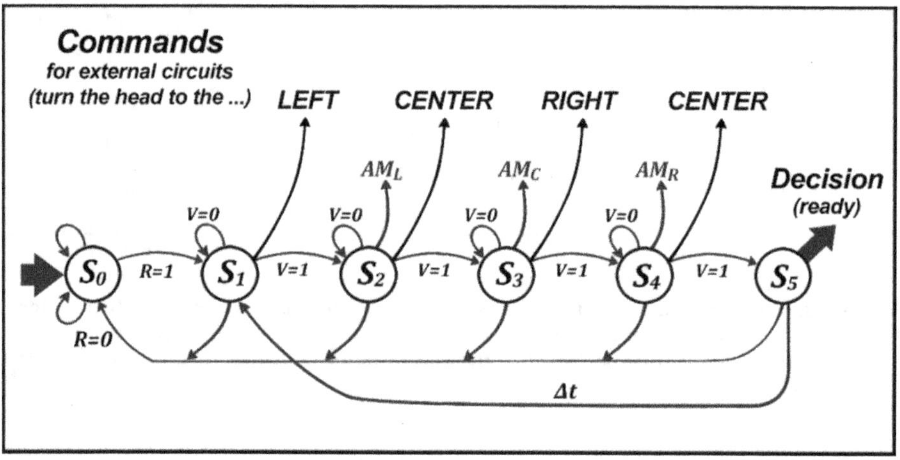

Figure 12.5: The searching-for-water state diagram. S_0 is the initial state, for the non-thirsty period. Assembly R triggers the algorithm execution by causing a transition to S_1. From S_1 to S_5 the agent repeats a loop that perform the algorithm. Please, refer to the text for detailed explanation.

We can use the threshold and hysteresis (T&H) network to decide when to *set* or to *reset* a behavior. For instance, it can be the starter for an algorithm we call seek-for-water behavior (SFW). During a non-thirsty period, few members in X fire; thus, the T&H output is OFF. However, in pre-thirsty and thirsty periods, the members of X increase their activities. Above a certain threshold, the agent triggers the SFW behavior, which may be like assembly R firing in the network shown in Fig. 12.4a. Thus, R triggers the SFW algorithm that controls the agent's behavior. In order to find water, the agent moves its legs, its head, besides performing other actions.

5.3 The algorithm in a FSA

For simplicity, we consider only few variables controlling the transitions in this finite-state machine: R is the output of a T&H block, indicating that the agent must keep searching for water. Variable Δt is a signal from an internal pacemaker. Variable V stands for a volition command coming from other blocks; thus, the agent moves its head when assembly V fires.

Three other variables in the algorithm are the output of greater of two (GoT) blocks, which compare the samples S_L, S_C, and S_R. As seen, the GoT can compare two analog samples at a time; the decision-making system must thus partially compare pairs of samples.

The routine starts at the S_0 state, meaning the agent is not thirsty. When R becomes active (R=1), the automaton goes to state S_1, starting the algorithm. From this point one, the agent is thirsty and it waits for volition commands for changing from one state to another.

We consider that, when in the S_1 state, the algorithm triggers a command to lower-level central pattern generators (CPGs) for turning the head left, then the FSA waits for another V.

When V fires again, the algorithm triggers a command to an *analog gate* (AG_L) for taking a sample of Y, the S_L sample, memorized in AM_L (analog memory for the left sample). The algorithm also sends another command to a CPG for turning the head to the center, and the FSA waits for another V. When the V event comes, the algorithm commands an AG (the AG_C) for sampling Y (the S_C), memorizing in AM_C. Simultaneously, the algorithm commands a CPG for turning the head right, and the FSA waits for another V. After a V, the algorithm triggers the AG_R for gathering in Y the S_R sample, stored in AM_R, and command a CPG for turning the head to the center again.

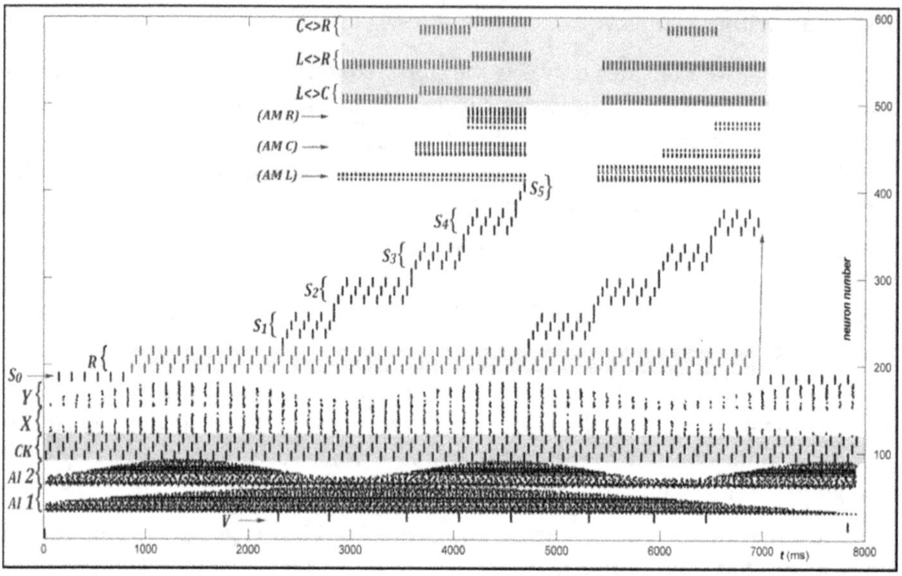

Figure 12.6: Raster plot of the algorithm shown in Fig. 12.5. Each assembly has 10 Leaky Integrate-and-Fire neurons. AI_1 and AI_2 are random analog inputs codified as population codes with 30 neurons each. Assembly *V* represents volitive input commands. For details, please refer to the text.

At this point, the automaton goes to the S_5 state. The state-machine uses this state for making the final decision, for controlling leg movements, and for going back to S_1. No matter which state the machine is in, when *R* becomes inactive (*R*=0), the automaton resets variables and states and makes the SFW behavior inactive. In other words, the network stops the execution of the seek-for-water algorithm.

5.4 The raster plot

Fig. 12.6a shows a raster plot for the algorithm simulated in Matlab with the Leaky Integrate-and-Fire (LI&F) neuron model. Sixty assemblies with *k* neurons each form the raster. From the bottom, the first three assemblies are reserved for introducing markers. The next are two analog input assemblies, each with $3k$ neurons each, which have the number of active members modulated by two sinusoidal signals. Assembly A_1 represents the thirst intensity, while A_2 represents the intensity perceived by an agent's humidity sensor.

The group of assemblies named *CK* forms an internal pacemaker. Whenever the lowest assembly in *CK* fires, it forces two AGs to take samples shown in Fig. 12.6 above the *CK* group. The *X* samples come from A_1, representing the agent's thirst intensity. The *Y* samples come from A_2, representing the intensity of humidity perceived by the agent's sensor.

A T&H topology receives the samples from *X*. While few members in *X* fire, the S_0 state is active, for example, from 0 to ≈800 ms or after ≈7000 ms. As the number of active members in *X* rises, the T&H network triggers *R*, which is a BNA that memorizes the thirsty state of the agent.

Since *R* is active, whenever the agent receives *volition* commands, the FSA changes to new states. These commands come from other modules and the raster shows only their events, the *V* assembly at the bottom of the raster. As *V* fires for the first time, the agent turns the head left, changing its internal state to S_1. As *V* fires again, the agent memorizes a *Y* sample (AM_L) and turns its head to the center, changing to the S_2 state (see Fig. 12.6 ≈ 2800 ms).

The raster represents three GoT networks above the analog memories (AM_L, AM_C, and AM_R). They perform GoT functions named: L<>C, L<>R, C<>R. L<>C compares the analog sample from AM_L with the sample in AM_C. At this point (during the S_2 state), there is nothing in AM_C, the L<>C output indicates that AM_L is greater than AM_C. The same happens to L<>R, because there is no sample memorized in AM_R.

When *V* fires again, AM_C memorizes an analog sample of the central position. The agent turns its head to the right. At this point (S_3), the GoT L<>C shows that AM_C is greater than AM_L. However, the GoT L<>R keeps indicating AM_L as the winner, because there is no sample in AM_R yet.

When *V* fires again, AM_R memorizes a sample when the head is in the right position. The agent's head turns back to the center. At this point (S_4 ≈ 4200 ms), all analog memories have samples.

The GoT L<>C shows that AM_C is greater than AM_L. The GoT L<>R indicates that AM_R is greater than AM_L; and finally, the GoT C<>R indicates AM_R is greater than AM_C. After another *V* event, the agent goes to the S_5 state. The agent uses these results (the GoT outputs) to command its legs and turn its body to the right. This is not shown in the raster because a simple combination of logic functions may account for the command that makes the agent perform this action. The outputs of such combinational logic become commands for concurrent algorithms, like what *V* does in this algorithm. From S_5, the algorithm changes to S_1 and the cycle is repeated.

At that point, the agent's body has turned a little aiming at a probable source of water, which it is looking for. In the new cycle (starting at ≈4600 ms), the AM_L sample, captured at ≈5500 ms, is greater than the AM_C sample, captured at ≈6000 ms. The AM_R sample, captured at ≈6500 ms, is smaller than the others. Hence, both L>C and L>R comparators remain active, indicating that the sample on the left is the winner.

This cycle is repeated until the thirst is quenched. At ≈7000 ms, the population firing for thirst becomes lower than 1/3 of the active input population, the T&H does not trigger the *R* loop, which goes to OFF. The network restores the S_0 state, dismantling the execution of the SFW algorithm.

This is a proof of concept showing how algorithms may become a behavior. It is well known that different algorithms can solve the same problem; thus, this may not be the best solution for seek-for-water problem.

6. Discussion

Let us recall the algorithm shown above. A similar algorithm for foraging was shown in Ranhel (2012). Note that NAC executes FSAs in parallel, hence, algorithms such as foraging, looking for a mate, escaping from predators, seeking a safe shelter, etc., are all concurrent and they may interact with one another. Hierarchical organization for algorithms must exist. If an agent is executing a search-for-mate and perceives a predator, then an escape routine must superimpose the algorithms in execution. A hierarchic control deactivates FSAs in favor of others that deal with crucial responses.

Some hierarchical controls may never change during the agent's lifetime; for example, the fight-or-flight response. However, considering intelligent and cognitive agents, many responses are modified as they undergo experiences, especially when the environment changes considerably. This means that our algorithm model must change. In other words, the algorithms must be created by adaptive automata, instead of hardwired finite-state automata.

Some important questions in this direction might be: how to create *adaptive* automata in NAC? How to create FSAs that learn how to modify themselves as they undergo experiences? Besides, how to create FSAs that learn how to construct other FSAs? Moreover, how can neurons co-opt and how can they discard members in assemblies?

Although bistable neural assemblies are able to create loops without changing synaptic weights, plasticity mechanisms in NAC are mandatory for modifying FSA. We are working on introducing spike-time dependent plasticity (STDP) in the SNNwD model as a mechanism that will be able to handle short-term modifications. The mechanism itself is not hard to be implemented; however, the STDP influence on the overall behavior is hard to determine. STDP operate in short time intervals (from –20 to +20 ms).

In order to deal with algorithms that change their behavior, let us say, after receiving a reward a few seconds later, we decided to introduce astrocytes in our SNNwD model. Astrocytes seem to play important roles in neuron interactions (De Pittà et al., 2015; Bazargani & Attwell, 2016). The timescale responses on astrocyte networks are slower than in the STDP mechanism. Unfortunately, this subject matter is extensive and is out of the scope of this chapter.

Considering all we have proposed above, the hypothesis of nervous systems composed of interacting CPGs (Yuste et al., 2005) is plausible with our proposals for NAC. STDP is considered the temporal Hebbian rule. This means that neural assemblies operating with Hebbian rules are closer to adaptive CPGs (adaptive automata).

Another discussion is about the world representation in our model. Physical phenomena are analog in their nature. In this sense, neural representations of analog phenomena are fundamental. Note that, for neural assembly computing (NAC), analog and digital domains operate seamlessly within the same substratum: spiking neural networks with delays. But, we showed here algorithms that use only rate- and population-coding schemes. As said earlier, a still open discussion in neuroscience is about how spiking neurons represent the world.

The neuroscience literature considers at least three groups of neural coding: firing-rate, population, and temporal coding. Within the temporal coding there are subcoding schemes (e.g. phase, ranking, latency coding). Temporal coding can represent huge amount of data (see VanRullen et al., 2005); thus, it is a promising way of computation and representation for spiking neural networks. However, the computing style for temporal coding is different from what we have developed herein.

Representation and functional blocks described above must be redesigned if we intend to work in the temporal coding domain. A still bigger challenge is to answer the following question: do firing rate, population and time-coding schemes operate together in nervous systems?

7. Conclusions

Cell assemblies naturally emerge in neural networks with action potential propagation delay. Delays cause spike coincidences in certain neurons; thus, sets of neurons fire simultaneously. In this scenario, neurons function much more as coincidence detectors than as integrators.

Representation and computation by cell assemblies mean an evolutionary advance, because groups of neurons are more robust and reliable than a single cell. Interactions among cell assemblies give rise to logical functions, decision-making, finite-state automata (FSA) and other functional blocks used for creating algorithms. Spiking neural networks (SNN) can execute many algorithms in parallel.

In animals, central pattern generators (CPG) are natural circuits that execute FSAs, which are responsible for gait generation, stereotyped responses, etc., culminating with complex reasoning and cognition. Hierarchically, FSAs may control CPGs in order to generate complex and intelligent behaviors. In this article, we have demonstrated how algorithms and complex behavior may emerge from relatively simple functional blocks created exclusively with spiking neurons. We show that our framework (neural assembly computing – NAC) is in accordance with the hypothesis of the neocortex being composed of interacting CPGs (Yuste et al., 2005). NAC can explain, from bottom-up, how SNN can generate intelligent and cognitive behavior by means of hierarchical organization of algorithms that interact with each other.

In the present stage of our research, we have been designing FSAs and algorithms. In fact, we are doing neural engineering. We have also been discussing how automata can modify themselves as they pass through experiences. This is a step towards constructing basal circuits that may become adaptive automata. Long-term potentiation (LTP) and depression (LTD) may change network topologies; thus, they may modify algorithms and automata in cell assemblies. In this sense, learning is associated to creating new FSAs or to modifying the existing ones. To deal with self-modification in algorithms, two models are being introduced in our framework: the STDP (spike-timing dependent plasticity) mechanism and a neuron-astrocytes interaction model.

Bibliography

Abeles, M. (1982). Role of cortical neuron: integrator and coincidence detector. *Israel Journal of Medical Sciences*, 18(1), 83-92.

Arbib, M.A. (ed.) (2003). *The handbook of brain theory and neural networks.* 2nd ed. Cambridge (MA): MIT Press.

Bazargani, N. & Attwell, D. (2016). Astrocyte calcium signaling: the third wave. *Nature neuroscience*, 19(2), 182-189.

Bekoff, M.; Allen, C. & Burghardt, G.M. (2002). *The cognitive animal: empirical and theoretical perspectives on animal cognition.* Cambridge (MA): Bradford/MIT Press.

Berwick, R.C.; Okanoya, K.; Beckers, G.J. & Bolhuis, J.J. (2011). Songs to syntax: the linguistics of birdsong. *Trends in Cognitive Sciences*, 15(3), 113-121.

Brette, R. (2012). Computing with neural synchrony. *PLOS Computational Biology*, 8(6), e1002561.

Buzsáki, G. & Draguhn, A. (2004). Neuronal oscillations in cortical networks. *Science*, 304 (5679), 1926-1929.

Cessac, B.; Paugam-Moisy, H. & Viéville, T. (2010). Overview of facts and issues about neural coding by spikes. *Journal of Physiology - Paris*, 104 (1-2), 5-18.

De Pittà, M.; Brunel, N. & Volterra, A. (2015). Astrocytes: orchestrating synaptic plasticity? *Neuroscience*, 323, 43-61.

Eliasmith, C. & Anderson, C.H. (2004). *Neural engineering: computation, representation, and dynamics in neurobiological systems.* Cambridge (MA): Bradford/MIT Press.

Hebb, D.O. (2002). *The organization of behavior: a neuropsychological theory.* Mahwah (NJ): Lawrence Erlbaum. Originally: 1949.

Hopcroft, J.E.; Motwani, R. & Ullman, J.D. (2007). *Introduction to automata theory, languages, and computation.* 3rd ed. Boston: Pearson/Addison Wesley.

Ijspeert, A.J. (2008). Central pattern generators for locomotion control in animals and robots: a review. *Neural Networks*, 21(4), 642-653.

Izhikevich, E.M.; Gally, J.A. & Edelman, G.M. (2004). Spike-timing dynamics of neuronal groups. *Cerebral Cortex*, 14(8), 933-944.

Kandel, E.R.; Schwartz, J.H. & Jessel, T.M. (2000). *Principles of neural science.* 4th ed. New York: McGrall-Hill.

König, P.; Engel, A.K. & Singer, W. (1996). Integrator or coincidence detector? The role of the cortical neuron revisited. *Trends in Neurosciences*, 19(4), 130-137.

Kumar, A.; Rotter, S. & Aertsen, A. (2010). Spiking activity propagation in neuronal networks: reconciling different perspectives on neural coding. *Nature Reviews Neuroscience*, 11(9), 615-627.

Paugam-Moisy, H. & Bohte, S. (2010). Computing with spiking neuron networks. In: Rozenberg, G.; Bäck, T. & Kok, J.N. (eds.). *Handbook of natural computing.* Heidelberg: Springer, pp. 1-47.

Purves, D.; Augustine, G.J.; Fitzpatrick, D.; Hall, W.C.; Lamantia, A.-S.; McNamara, J.O. & Williams, S.M. (eds.) (2004). *Neuroscience.* 3rd. ed. Sunderland (MA): Sinauer.

Ranhel, J. (2012). Neural assembly computing. *IEEE Transactions on Neural Networks and Learning Systems*, 23(6), 916-927. Available at: http://ieeexplore.ieee.org/lpdocs/epic03/wrapper.htm?arnumber=6186825.

——— (2013). Neural assemblies and finite state automata. In: Engelbrecht, A.; Bastos Filho, C.J.A. & Lima Neto, F.B. (eds.). *Proceedings of the 1st BRICS Countries Congress on Computational Intelligence (BRICS-CCI)*. Los Alamitos (CA): IEEE Computer Society, pp. 28-33. Available at: http://www.neuralassembly.org/download/NAC_Finite_State_Automata_BRICS-CCI-2013.pdf.

Smith, C.U.M. (2000). *Biology of sensory systems*. Chichester (GB): John Wiley & Sons.

VanRullen, R.; Guyonneau, R. & Thorpe, S.J. (2005). Spike times make sense. *Trends in Neurosciences*, 28(1), 1-4.

Wickens, J.R. & Miller, R. (1997). A formalisation of the neural assembly concept. 1. Constraints on neural assembly size. *Biological Cybernetics*, 77(5), 351-358.

Yuste, R.; MacLean, J.N.; Smith, J. & Lansner, A. (2005). The cortex as a central pattern generator. *Nature Reviews Neuroscience*, 6(6), 477-483.

Chapter 13

Language, communication, and cognition

André Leclerc
Institute of Human Sciences
University of Brasília, Brasília, DF, Brazil
andre.leclerc55@gmail.com

William A. Pickering
Piracicaba School of Technology (FATEP), Piracicaba, SP, Brazil
willpickering@yahoo.com

Abstract

In the first part, André Leclerc argues that concepts proposed by philosophers of language, such as truth conditions and the recognition of speaker intentions, are not enough for explaining linguistic meaning. Instead it is proposed that what gives a person access to intuitive truth conditions, those immediately grasped by the agents of the context, is a previous understanding of the whole situation, in particular a grasping of speakers-hearers' expectations and plans. In the second part, William Pickering summarizes the relevance of complex system theory to linguistics. It is argued that concepts from systems theory such as distributed control, feedback, emergence, and reciprocal causality can be used to understand the relationship between language system and language change, as well as between language and thought.

Introduction

This paper presents two critical views on current theories of language, communication, and cognition. Section 1 criticizes traditional concepts in the philosophy of language as they relate to linguistic meaning, and proposes instead that a listener's knowledge of immediate situational context, including a grasping of the expectations and plans of a speaker, is necessary for linguistic understanding. Section 2 gives a general overview of the relevance of complex system theory to the study of human language, with reference to the most important recent books and articles on the subject.

1. Spontaneous linguistic understanding

If you think for one moment of a typical situation of human communication, what will come to your mind is probably the following scene: two or more people, knowing each other for some time and speaking their mother tongue, perform acts of utterance with the intention to be understood or to cause some other effects in the mind of the other agents of the context. A contrast could be usefully established with two very different experiences: when you do not master fully the language used in the communication (interlinguistic communication), or when you have to interpret because what has been said is not so clear at first sight, or has been said with a certain degree of sophistication that involves inferences and reflection (for more details, see Leclerc, 2012). Interlinguistic communication and hermeneutical practices are uncommon and not at all immediate or spontaneous. For most people on Earth, spontaneous linguistic understanding (SLU) is by far the most common experience of communication. Philosophers of language use a battery of concepts to describe that very common situation and to explain the success of communication.

1.1 Disquotational truth conditions

But first, we must have a look at a very powerful concept by which philosophers, for more than a century, explain what it is for someone to understand a sentence: the concept of *truth conditions*. Introduced by Frege ([1893] 1964), it gives an elegant answer to the question: what is it to have a thought? To have a thought is the same as thinking that certain conditions for its truth are fulfilled. Wittgenstein ([1921] 1960) gave the next step: to understand a declarative sentence is to know how the world must be if the sentence is true; in other words, it is the same as representing its truth conditions. The understanding of a sentence and its truth conditions are conceptually linked. All the tradition of truth-conditional semantics is based on that simple idea. To understand a sentence is to construct somehow a representation of its truth conditions.

Tarski ([1933] 1956) used this idea when he provided a definition of truth for an important class of formal languages. The semantics of a formal language L must generate recursively, for each sentence of L, a theorem (call it a "T-Sentence") of the form:

(T): S is true in L if and only if P,

where on the left side of 'if and only if' we have an application of the truth predicate to the name ('S') of a sentence, and on the right side a description of the truth conditions of that sentence. This is the famous "T-Convention", well-known to any student of logical semantics. What we understand is meaning and meaning is analyzed in terms of truth conditions. To master a language, therefore, is to be able to form a T-Sentence for any declarative sentence of the language ('Snow is white' is true in English if and only if snow is white; 'Ottawa is the capital of Canada' is true in English if and only if Ottawa is the capital of Canada, etc.). These are "disquotational" truth conditions, because the T-Sentence says that both sides of "if and only if" are equivalent; therefore, erasing the truth-predicate and removing the quotation marks yields the same as the right side.

That approach works well for formal languages, or for sentences of natural languages with no indexicals, no demonstratives, no context-dependency. But as a description of understanding in natural languages, it is seriously wanting. Things are not so straightforward. First, it is not true that a standard T-Sentence always gives us all we have to know in order to understand a sentence. Our knowledge of nature is as limited as our knowledge of language. A sentence like 'There is life on Mars' (Dummett's example) can have more than one understanding. How must the world be if that sentence is true? It is far from clear, and the understanding may be quite different according to the background knowledge of different people (a NASA specialist in exobiology will have a better understanding than we do on such a matter.) So, a T-Sentence, like "'There is life on Mars' is true in English if and only if there is life on Mars", looks quite uninformative and trivial (Barber, 2003). Merely disquotational truth conditions give us, in many cases, a very poor understanding of what it is to understand a sentence.

Second, context-dependency is quite pervasive in natural languages. It is probably the most important topic in philosophy of language since the works of Russell (1940) and Reichenbach (1947). As McGinn (2000, p. 46) once nicely said, "indexicality cannot be captured non-indexically." The idea that for any sentence containing an indexical (or a demonstrative) there is a corresponding sentence, a paraphrase with the same sense, from which the indexical has been removed, is false.

Davidson (1967) believed that, with some adjustments, it is possible to account for the "demonstrative aspects" of natural languages, by adopting a version of Tarski's T-Convention for natural languages. Our semantic theory for English, for instance, would thus entail T-sentences of the form:

'I am thirsty' is true as (potentially) spoken by *p* at *t* if and only if *p* is thirsty at *t*.

'That bird is a greenfinch' is true as (potentially) spoken by *p* at *t* if and only if the bird demonstrated by *p* at *t* is a greenfinch.

Etc.

In these schematic sentences, *p* is the speaker or agent of the context, and *t* the time of the utterance. We are now considering sentences uttered by an agent at a time. But the identity of the speaker and the time of the utterance are not enough. We need many more parameters, and it is not clear at all how we could close the list. Suppose you are talking from Brazil to a friend living in Sweden and that friend says "I am going to spend the winter with you in Brasília." To understand the sentence correctly you have to establish clearly what winter he has in mind: austral or boreal?

The philosophy of language exposed in Wittgenstein ([1953] 2009) and Austin ([1961] 1979) shows how hard it is to separate semantic knowledge from the rest of our encyclopedic knowledge, and how complex are the things we do with linguistic expressions. In other words, our understanding of sentences depends most of the time on so many contextual factors that it cannot be captured easily by a Tarski-Davidson type of semantics (model-theoretical), even in the rigorous Montague version (1970).

A different program, that of Grice (1957, 1975), explains our understanding of sentences in terms of recognition of speaker's intentions. Suppose I see you on the street and wave my hand in a typical way with the intention to salute you; but you do not look in my direction. So I did not salute you at all. But suddenly you see me waving my hand again and recognize immediately my intention to salute you; only then I succeed. As long as the speaker's intentions are not recognized, the communication fails. Human communication would be an affair of making manifest our intentions to communicate something, and of recognizing these intentions. But Grice's program leaves the semantics of "what is said" (the content of the utterance) more or less on the same footing: what is said is still the result of semantic conventions (which determine standing or lexical meanings) and the "saturation" of demonstratives and indexicals (when elements of the context are associated with 'I', 'you', 'now', 'yesterday', 'here', 'this', etc.). In their programs, when it comes to context-dependency, Davidson and Grice do not go beyond the saturation of indexicals and demonstratives. So, the understanding of a new sentence depends only upon the knowledge of lexical (standing) meanings and the way the lexical elements are combined, plus the knowledge of the contextual elements needed for the saturation of indexicals and demonstratives. Nothing more. In that perspective, semantics is insensitive to contextual influence; with the exception of demonstratives and indexicals, the rest of our semantics is not "occasion-sensitive."

1.2 Intuitive truth conditions

Consider the following sentence-types:

a) Jones had a walk.

b) Jones has got the virus.

c) There is a lot of coffee on the table.

Different tokens of A may have very different truth conditions, for example, when 1) Jones is an adult trying to stay in shape; 2) if Jones is a toddler that just gave his first steps in his life; 3) when Jones is an athlete that has just undergone a surgery in his knee; or 4) when Jones is an elderly person in a hospital recovering from a serious disease. The situations in which different tokens of A can be *literally* true are very different from one another. And each utterance of A must be understood differently to be understood correctly. It does not seem possible to close the list. The only limit here is our imagination. The sentence-type just determines a class of possible situations in which tokens produced in different utterances of A could be literally true. It is now easy to see that the T-Sentence "'Jones had a walk' is true if and only if Jones has a walk" is totally uninformative. The same holds for "Jones has got the virus": in one situation, Jones is in his bed, with fever, taking medicines, etc.; in another, Jones is an epidemiologist that just got a sample of a dangerous virus. Now imagine a big pot of coffee on a table; an utterance of C is literally true and could be interpreted as an indirect invitation to serve oneself; now, uttered by the boss of an industry that grinds coffee beans in a room where there are a lot of bags of coffee beans on a table, the assertion of C, again, will be literally true, and could be understood as an indirect order to do something with the bags; suppose someone spills a whole cup of coffee on a table; again, an utterance of C would be literally true, and could be understood as an indirect request to clean the table. Undoubtedly, we have the ability to understand differently different tokens of the same type according to the context. So *our understanding of the situation contributes in a decisive way to the correct understanding of a token produced in a specific context of use.* What gives us access to the intuitive truth conditions, those immediately grasped by the agents of the context, is a previous understanding of the whole situation, *in particular a grasping of their expectations and plans.*

But there is more. We understand easily ungrammatical sentences, nonsense, malapropisms, incomplete sentences. Consider the following sentences:

a) You are very much Paris.

b) Give peace a chance!

c) Pride has a city!

d) The Cardinal Mazarin has sent here his hemispheres.

e) I am ready!

f) John speaks fluently English.

The first three are nonsenses, but they are easily understood; (d) is a malapropism (just switch 'hemispheres' for 'emissaries'); (e) is incomplete (ready for what?) but understood easily on each occasion of use; and the last one is perceived as ungrammatical (the adverb should go to the end of the sentence).

These cases are marginal, of course, but they reveal something interesting for my purpose. Spontaneous linguistic understanding (SLU) is understanding of utterances, of actions performed for such and such a reason, and its first and primary object is occasion meaning. Occasion meaning is the result of a pre-propositional "modulation" (Recanati's 2004 expression) of standing meaning. The common use of language is a rule-governed rational activity, and rationality is precisely an unlimited capacity to realize intentional adjustments in a huge diversity of contexts. Linguistic understanding goes much beyond the mere mechanical parsing of sentences. It presupposes much more. Here is my working hypothesis:

> *Actions and practices are the roots of linguistic understanding. [...] Words and sentences are seen and understood as instruments used in the performance of actions. My suggestion is that we should consider plans and not only actions, as the unit of investigation. I also suggest that the expectations of agents (speakers-hearers), and the correct identification and understanding of these expectations, are especially important for the determination of the sense of the words used in an occasion. In other words: The root of any distinction in thought and in the sense of linguistic expressions is found in its sensible effects, in our practices, plans and activities. This principle I call the Principle of the Determination of Sense.* (Leclerc, 2009, p. 271)

In the view here defended, SLU *is not* modular (see Borg, 2004, for the opposite position); it uses information of different sources and its domain is not well specified, so that the two main criteria for modularity (informational encapsulation and domain specificity) are not satisfied. Our actions are parts of plans, and this holds for speech acts too. We are obliged all the time to realize adjustments in an infinity of different situations. To be successful most of the time in these processes of adjustments is what makes us rational agents. The abilities to speak and understand discourse is just part of that larger process.

2. Linguistics and complex systems theory

The application of complex systems theory in linguistics is relatively recent, although interest in applying complex systems theory to language-related phenomena began with some of the founders of cybernetics, information theory, and dynamic systems theory. Andrei Kolmogorov, Claude Shannon, and Benoît Mandelbrot, for example, researched from a mathematical point of view how concepts such as feedback, complexity, and information relate to human language. This work never gained influence in mainstream linguistics, despite its extensive development by researchers in the field of quantitative linguistics (Köhler 2005; Köhler & Altmann, 2005). Within the last twenty years, however, an increasing number of publications have appeared that deal with one or another aspect of language from the complexity perspective (the most important of these are reviewed in Kretzschmar, 2015). There cannot rightly be said to exist at present a unified theory of language derived from the complex systems viewpoint, but rather a general perspective, adopted by scholars from different linguistic subfields, that pretends to approach linguistic phenomena on the basis of certain principles taken from complexity theory. A brief overview of the basic concepts of this theoretical perspective is given below.

How does the complex systems approach relate to the study of language, and how does this approach, when applied in linguistics, relate to cognitive science? First of all, there are several important characteristics that natural languages share with complex systems (Beckner et al., 2009; Kretzschmar, 2009, 2015; Larsen-Freeman & Cameron, 2008). In general, complex systems are composed of units which are in dynamic interaction, and it is the interdependency of these units in interaction that gives the system as a whole a functional character. Each unit interacts with other units on the local level, but this local interaction sustains the larger system as a functioning whole. One can therefore conceive of languages as dynamic systems, and individual speakers as units in the system.

Conversation is a form of interaction, and it is a local activity in which speakers' actions are highly interdependent. It can also be considered on various timescales. One can think of it on a micro-scale in terms of the activity of the brain, or, on a more extended timescale, in terms of the immediate conversational setting. Or one can expand the timescale and think in terms of the types of personal relationships that speakers have over time, and on a larger timescale one can even think of cultural processes, or of the multigenerational processes of language change.

The interaction of the units in a complex system can be defined in terms of a network structure (Heylighen, 2008), and social relationships can be described as network structures (Easley & Kleinberg, 2010). The important thing in the context of the present discussion is that human communication is essential in the formation of these networks. Sociolinguists have found that sociolinguistic variance (roughly speaking, dialect differences related to speaker characteristics such as age, gender, and social class) correlate with social networks, and recently this phenomenon has been the subject computational modeling (e.g., Fagyal et al., 2010; Mühlenbernd & Quinley, 2013).

Another important aspect of complex systems is distributed control, which means that there is no external or central controller, or single unit or group of units in the system, that directs the actions of the system. Languages are a good example of this kind of decentralized control, because nobody actually legislates how people should talk. Language academies or school systems may enforce standards for written language with a certain degree of success, but they cannot stop the inevitable changes that occur in spoken language. An important aspect of human language, therefore, is that it is a kind of spontaneously evolved social norm. How this comes about is something that needs to be explained.

Probably the key concept of the complex systems perspective is the idea of *emergence*. Emergent properties are properties of a complex system that cannot be reduced or explained only in terms of an analysis of the individual units of the system, but that come into being, or emerge, as a result of the interaction of the system's units. This is very important, because based on this idea it may be concluded that when one talks about thought, or cognition, or the mind, one must take into account how individual characteristics come about as part of a larger social process, and something essential will be missing if one tries to isolate an explanation of language or cognition from an explanation of how humans interact with each other.

The core hypothesis of the complex systems approach to language has been well summed up by Beckner et al. (2009, p. 15), when they say that "An idiolect is emergent from an individual's language use through social interactions with other individuals in the communal language, whereas a communal language is emergent as the result of the interaction of the idiolects". It should be noted that computer modeling is useful in this kind of study, because it is very difficult to study how large numbers of people interact over long periods of time. Beckner et al. (2009, p. 12) point out that computer modeling "allows one to prove, at least in principle, that specific fundamental mechanisms can combine to produce some observed effect".

Although complex systems exist as a result of, or one could say are caused by, the individual activities of the units in a system, the emergent properties – once they have emerged – can also restrict the behavior of those units. A simple example of this is the fact that once a person becomes part of a language community, that individual is in a sense obligated to use certain words. A person is not required to follow the norm, but if that person doesn't, he or she is not going to get anything done. The individual's behavior has therefore been restricted by the larger system – one must go along to get along. This is a two-way process, however, which is sometimes referred to as "reciprocal causation" (Larsen-Freeman & Cameron, 2008, p. 59). On the one hand, individuals can say what they like and can create innovations. These innovations can be copied and spread from person-to-person, "causing" language change (i.e., changes in the system) from the bottom up. On the other hand, as noted earlier, the system restricts the individual's behavior at the same time. Thus, the system, from the top down, "causes" the individual to conform.

Another main characteristic of complex systems are feedback processes. If we think of a system in network terms, negative feedback within a network tampers perturbations, and this allows the system to have stability. Positive feedback can have the opposite effect, reinforcing these perturbations so that they can reverberate through the system and bring about change. In complex systems, positive and negative feedback interact to allow a system to maintain stability, while at the same time giving it flexibility to respond to changes in the environment. In terms of language, the entry into a language of an innovation such as a new word, grammatical construction, or way of pronouncing something, can be thought of a process involving positive and negative feedback in a network of speakers. For example, some innovations are picked up and repeatedly imitated, which would be a case of positive feedback. To take an extreme example, the word "gas" was invented by a single individual at the beginning of the 17th century (Bloomfield, 1914, p. 237). An innovation may be imitated by only a relatively small group, such as family or a clique, and might fade into disuse after a relatively short time (negative feedback), as is often the case with slang.

The above discussion contains just a few of the analogies that can be made between complex systems and languages, and many others could be made to fill out this picture. Within linguistics at present, most scholars who adopt the complex systems approach do so from what is called the usage-based perspective. The usage-based perspective is in fact a rather important school of thought within linguistics, and in essence what it proposes are explanations of things like language acquisition, grammatical structure, and language change, that do not depend on the assumption of highly specific and restrictive language capacities in the brain. For Bybee (2008, p. 233), "a great advantage of Usage-Based Theory over Generative Theories is that it does not rely on innateness to explain linguistic categories but rather proposes that much of grammar can be explained on the basis of the domain-general abilities of humans", such as analogy, categorization, chunking, and automatization. (This is, of course, in opposition to the formalist or innatist school of thought, also called generativism, whose main representative is Noam Chomsky.) What the complex systems perspective adds is to show how it is possible, through the collective interactive activity of individuals, for a communal language to come into being – that is, to *emerge* in the sense that complex systems theory gives to this term – and to maintain its stability as a norm of communication while also changing over time. In regard to the relevance of usage-based theory to the complex systems approach, the Spanish linguist Enrique Bernárdez (2008, p. 157-158) has made the interesting observation that many of general capacities of the mind that are relevant to language, such as imitation and the ability to identify the intentions of others, are directly related to social interaction.

In conclusion, the above argument amounts to more than just stating the obvious fact that people are products of their environment. The complex systems perspective on human languages also goes beyond the idea, which has long been argued for in many fields outside of linguistics, that the specific forms of language and cognition that we observe in individuals cannot be understood properly if they are isolated, or separated, from interactive activity such as conversation. The complex systems perspective is a way of explaining, through concepts such as feedback, emergence, networks, and reciprocal causality, how individual language behavior is formed by the social environment, and how at the same time the individual contributes to the collective creation and maintenance of social norms of language and cognition. Furthermore, it provides a way of understanding how local processes, such as conversation, are connected to larger social processes such as sociolinguistic variation and language change. Most importantly, for the purposes of this discussion, it shows that an understanding of the emergence of language through interaction is essential to any account of how the mind works.[1]

[1] Section 1 was written by André Leclerc, and section 2 by William A. Pickering.

Bibliography

Austin, J.L. ([1961] 1979). *Philosophical papers.* Eds. J.O. Urmson & G.J. Warnock. 3rd ed. Oxford: Clarendon.

Barber, A. (2003). Truth conditions and their recognition. In: Barber, A. (ed.). *Epistemology of language.* Oxford: Oxford University Press, pp. 367-395.

Beckner, C.; Blythe, R.; Bybee, J.; Christiansen, M.H.; Croft, W.; Ellis, N.C.; Holland, J.; Ke, J.; Larsen-Freeman, D. & Schoenemann, T. (2009), Language is a complex adaptive system: position paper. *Language Learning,* 59 (suppl. 1), 1-26.

Bernárdez, E. (2008). Collective cognition and individual activity: variation, language and culture. In: Frank, R.M.; Dirven, R.; Ziemke, T. & Bernárdez, E. (eds.). *Body, language and mind.* Vol.2: *Sociocultural situatedness.* Berlin: Mouton de Gruyter, pp. 137-166.

Bloomfield, L. (1914). *An introduction to the study of language.* New York: Henry Holt.

Borg, E. (2004). *Minimal semantics.* Oxford: Oxford University Press.

Bybee, J. (2008). Usage-based grammar and second language acquisition. In: Robinson, P. & Ellis, N.C. (eds.). *Handbook of cognitive linguistics and second language acquisition.* New York: Routledge, pp. 216-236.

Davidson, D. (1967). Truth and meaning. *Synthese,* 17(3), 304-323.

Easley & Kleinberg, J. (2010). *Networks, crowds, and markets.* Cambridge (GB): Cambridge University Press.

Fagyal, Z.; Swarup, S.; Escobar, A.M.; Gasser, L. & Lakkaraju, K. (2010). Centers, peripheries, and popularity: the emergence of norms in simulated networks of linguistic influence. *University of Pennsylvania Working Papers in Linguistics,* 15(2), 81-90.

Frege, G. ([1893] 1964). *The basic laws of arithmetic.* Transl. M. Furth. Berkeley: University of California Press.

Grice, H.P. (1957). Meaning. *Philosophical Review,* 66(3), 377-388. Reprinted in Grice, P. (1989), *Studies in the way of words.* Cambridge (MA): Harvard University Press, pp. 213-223.

——— (1975). Logic and conversation. In: Cole, P. & Morgan, J. (eds.). *Syntax and semantics.* Vol. 3: *Speech acts.* New York: Academic Press, pp. 41-58. Reprinted in Grice, P. (1989), *Studies in the way of words.* Cambridge (MA): Harvard University Press, pp. 22-40.

Heylighen, F. (2008). Complexity and self-organization. In: Bates, M.J. & Maack, M.N. (eds.). *Encyclopedia of library and information sciences.* Vol. 2. Oxford: Taylor & Francis, pp. 1215-1224.

Köhler, R. (2005). Synergetic linguistics. In: Köhler, R.; Altmann, G. & Piotrowski, R.G. (eds.). *Quantitative Linguistik/Quantitative linguistics: ein internationales Handbuch/an international handbook.* Berlin: Walter de Gruyter, pp. 760-774.

Köhler, R. & Altmann, G. (2005). Aims and methods of quantitative linguistics. In: Altmann, G.; Levickij, V. & Perebejnos, V. (eds.). *Problemy quantitativnoj lingvistiki.* Černivci (UA): RUTA, pp. 12-41.

Kretzschmar, W.A. (2009). *The linguistics of speech.* Cambridge (GB): Cambridge University Press.

——— (2015). *Language and complex systems.* Cambridge (GB): Cambridge University Press.

Larsen-Freeman, D. & Cameron, L. (2008). *Complex systems and applied linguistics.* Oxford: Oxford University Press.

Leclerc, A. (2009). Meanings, actions and agreements. *Manuscrito,* 32(1), 249-282. Special issue: Leclerc, A.; Perini-Santos, E. & Ruffino, M. (eds.). *Semantic content and communication.*

——— (2012). Spontaneous linguistic understanding: a few introductory remarks. *Disputatio,* 4(34), 713-738.

McGinn, C. (2000). *Logical properties.* Oxford: Oxford University Press.

Montague, R. (1970). Universal grammar. *Theoria,* 36(3), 373-398. Reprinted in Thomason, R.H. (ed.). *Formal philosophy: selected papers of Richard Montague.* New Haven: Yale University Press, 1974, pp. 222-246.

Mühlenbernd, R. & Quinley, J. (2013). Signaling and simulations in sociolinguistics. *University of Pennsylvania Working Papers in Linguistics,* 19 (1), 129-138.

Recanati, F. (2004). *Literal meaning.* Cambridge (GB): Cambridge University Press.

Reichenbach, H. (1947). *Elements of symbolic logic.* New York: Macmillan.

Russell, B. (1940). *An inquiry into meaning and truth.* London: George Allen & Unwin.

Tarski, A. ([1933] 1956). The concept of truth in formalized languages. Transl. (from the 1935 German translation from Polish) by J.H. Woodger. In: Tarski, A. *Logic, semantics, metamathematics.* Oxford: Oxford University Press, pp. 152-278.

Wittgenstein, L. ([1921] 1960). Transl. C.K. Ogden. *Tractatus logico-philosophicus.* London: Routledge & Kegan Paul.

——— ([1953] 2009) *Philosophical investigations.* 4th ed. Transl. G.E.M. Anscombe, P.M.S. Hacker & J. Schulte. Oxford: Wiley-Blackwell.

Chapter 14

Cognition in Peirce's semiotic

Ivo Assad Ibri
Graduate Studies Program in Philosophy
Pontifícal Catholic University of São Paulo (PUC-SP), São Paulo, SP, Brazil
ibri@uol.com.br

João M. Queiroz
Institute of Arts and Design
Federal University of Juiz de Fora (UFJF), Juiz de Fora, MG, Brazil
queirozj@gmail.com

Vinicius Romanini
School of Communication and Arts
University of São Paulo (USP), São Paulo, SP, Brazil
viniroma@gmail.com

Abstract

This paper presents answers to three different questions pertaining to the philosophy of Peirce, semiotics, and cognition. In section 1, after reviewing the three categories of Peirce's phenomenology, it is argued that pragmatism is an indispensable framework for semiotics. Section 2 explores further aspects of Peircian epistemology, pointing out the relevance of the semiotic theory of mind for current Cognitive Science. Section 3 examines Peirce's concept of information, showing how it is related to cognition.

1. Is pragmatism a necessary framework for semiotics?

In order to answer the question of whether pragmatism is a necessary framework for semiotics, we should start by remembering that the classification of sciences, according to Charles S. Peirce (1839-1914), establishes that Phenomenology is the first science of Philosophy (see, for instance, CP 1.280, CP 1.284, CP 5.121). Semiotics, in turn, is the third science of the Normative Sciences, after Esthetics and Ethics, occupying the place of Logic, with which it maintains a relationship of equivalence (see CP 5.129, CP 2.197, and, fundamentally, CP 2.227). The ranking of the sciences as proposed by Peirce is hierarchically structured, namely, based on an ordered dependence, such that the Normative Science is dependent on Phenomenology, and Semiotics is dependent on Esthetics and Ethics, so that all of these three are dependent on Phenomenology. Metaphysics completes the third class of sciences (see, for instance, CP 5.121, EP 2.259) belonging to Philosophy, as is well known by Peircean scholars.

Let us also recall that Phenomenology is the science of phenomena, namely, of everything that may appear to our senses, in the inner or outer worlds, and that may be a matter of experience (see EP 2.259, and also Ibri, 2015, ch. 1). This term "experience", by the way, has a special meaning in Peirce's vocabulary, bringing the idea of what can be added to a learning process as the course of life, in his own words.

Phenomenology will found three modes through which such experience will run. They are, supposedly, universal modes of being, which Peirce calls "categories". These categories, so well-known to Peircean scholars, will be the main conceptual axis for reading not only the modes of experience, but also the modes on which reality is constituted (see EP 2.179-195).

Assuming it will be easier to begin the exposition of the categories by the second one, called "secondness", one may say that it is the mode in which we experience the reaction from the world external to us, which shows itself not submitted to our will, neither running according to our predictions. This sort of reaction occurs through something that is definitely determined, which suggests itself as a hard-fact because its occurrence appears independent of our will or prediction. "Reaction" and "hard-fact" are a same meaning duet and almost synonymous. The secondness' main characteristic is to be the mode by which an external world appears to every mind supposedly able to interpret its composing facts. Secondness might be considered the category open to every cognitive mind. Some facts will be apprehended as regularities and some not. The former will be object of general representations, within the realistic hypothesis that they are replicas of laws, or general logical and somehow real stable logical relations. This hypothesis is ontological in its nature and derives from what phenomenological experience suggests. Reality will be composed by facts as an open and external mode of being called secondness, and by logical and continuous relations under them, which are called by Peirce "Laws" (see CP 5.1.421, CP 5.107, EP 2.67-74).

Laws will compose the third category, "thirdness", on its ontological face, based on the phenomenological experience of regularity that allows representing the facts as feasible to be thought. Thought, while mediation related to represented hard-facts, and law are then connected as belonging to the same category.

Finally, let us consider the first category, called "firstness". It is maybe the most difficult to be understood, as it deals with formless phenomena regarding logical rules (see Ibri, 2010; 2015, pp. 29-35). Spontaneity is a fair predicate for the phenomena that are under it. These phenomena appear in immediate interpretants, that mainly arise as "qualisigns", like sounds, colors, smells, that according to Peirce may be experienced as qualities of feelings. Under the genuine root of Peircean realism, qualities have ontological dimension and correspond to feelings, in the sense of sharing the same nature (see CP 6.198, CP 7.530, CP 1.531, EP 2.160, and CP 2.254, for an application in Semiotics). Peirce's cosmology is an exquisite theoretical building that tries to justify such sameness by conceiving the origin of categories, dealing with continuities of possibilities and of logical order. Indeed, it is according to him a conjecture, an abduction that will found what could be called a "genetic philosophy", a sort of reflection that radically refuses any kind of dogmatism, which often can be found in philosophical frames that cannot deal with originary signs, both in the sphere of theory and in ontology.[1]

Firstness will appear in its ostensive side in irregularities, as facts that cannot be generalized at all. These irregular, asymmetrical features of facts cannot be explained by natural laws, whose replicas exhibit regular predicates, sameness, redundancy, symmetry, everything that cannot be found in the firstness replicas. On the ontological side, while regular replicas are due to law, the irregular ones must be attributed to a different principle, which is Chance, according to Peirce (see, for instance, W 8.101, W 8.110, W8.120, CP 1.407).

With these considerations, it seems licit to say that secondness is the theater of reactions where these two kinds of replicas appear and show themselves as the external side of what can only be known by inferential means, namely the other two categories, firstness and thirdness. Actually, the specific characteristics of these replicas are somehow mixed as factual phenomena, i.e., laws are of a probabilistic nature, meaning that they include some degree of order, logical rules, and some degree of deviation in relation to average parameters.

Still under a realistic philosophical approach, every science can be understood as a semiotic activity, properly called semiosis, as the signic process for generating knowledge. They must interpret those replicas as the only accessible face of laws – actually, their external face. These replicas, signs indeed, are full of meaning and point to rules of conduct, habits of acting, we might say, to general signs, laws that, in the end, are the target of every science. We also might fairly say: indexical signs as replicas are consequences[2] of symbolic signs as laws. Semiosis (see Romanini, 2014), as a scientific process, is then constituted by an inferential knowledge of general signs (see Ibri, 2011) through particular signs, namely, signs

---

[1] The rejection of dogmatisms based on Peirce's realism is discussed in Ibri (2012).

[2] This term is quite meaningful, due to the following considerations about Pragmatism in its relationship to Semiotics.

open to direct experience in secondness are the way through which thirdness can be accessed in order to discover symbols as laws.

Let us now remember the conceptual axis of Peirce's pragmatism. It asserts that the meaning of a concept or general sign is the conceivable set of consequences that would be able to affect conduct (see Ibri, 2015, ch. 6). Originally concerned with concepts, the mature realism of Peircean philosophy will allow interpreting the pragmatism maxim in a broader sense, that of an ontological nature. Such amplified meaning would be a consequence of an ontological reading of the categories, following the former considerations here exposed: secondness is the external side, the consequences of the other two categories, which are only accessible by inference, as Law and Chance may be.

If we see human actions as consequences of concepts, as their external side, then it seems evident that there is a categorial similarity between natural facts and human actions, as both are, in their actuality and particularity, indexical replicas, through which we can semiotically interpret generals, whether they are laws or concepts. That is why we may consider the idea of habit as a heuristic unifying name for both, which somehow tacitly includes that similarity.

Our interpretants result in conduct, as well as all beings in the world express the way they interpret their specific environment, though the way they manifest themselves existentially, namely as visible and meaningful conduct, is able to be interpreted and known.

Pragmatism, on its ontological face, surpassing the mere original rule for distinguishing concepts' meanings, actually supports Semiotics by understanding modes of acting, conduct, as interpretants. Actions and conduct are signs. They may be known directly or inferentially, and this epistemological condition occurs through the ontological relationship between the outer and inner worlds, where the former becomes the way to access the latter. This relationship is, in synthesis, the main core of pragmatism.

A realistic philosophy will allow claiming that "everything speaks" – their language assumes a manifold of signical forms, among which the practical consequences, present in the classical Peircean pragmatic maxim, shall be recognized as interpretants of every being in the world, as well as matter of interpretation by any knower mind.

Pragmatism is a necessary framework for Semiotics as the indispensable way to insert this science of signs in Peirce's whole philosophical system, which will give it a wider capacity for understanding signs and their meanings, inside a context of a radical realism that is the proper essence of the Peircean mature thought. By interlacing both doctrines, under the bond of necessity, we will be able to understand Semiotics beyond the narrow conception of a mere taxonomic science of signs.

2. How can semiotics be applied now-a-days in the cognitive sciences?

Peirce can be considered an important precursor of the situated mind and distributed cognition thesis. In the Peircean Semiotic Theory of Mind, the fundamental unit of cognitive interest is reconceived – disembodied mind is replaced by the environmentally embedded space of semiotic skills and artifacts. According to the Peircean model, a meaning process involves a relational complex constituted by three irreducibly connected terms – Sign, Object and Interpretant (S-O-I) (see Fig. 14.1).

The irreducibility indicates a logical property of this complex: the sign process must be regarded as associated to the interpretant, as an ongoing process of interpretation (Hausman, 1993, p. 9), and is not decomposable into any simpler relation (CP 5.484). Peirce also defines a sign as a medium for the communication of a form or habit embodied in the object (De Tienne, 2003; Bergman, 2000). This form is communicated to the interpretant, so as to constrain (in general) the interpretant as a sign or (in biological systems) the interpreter's behavior.

The object of sign transmission is a habit (a regularity, a rule of action, or a 'pattern of constraints') embodied as a constraining factor of interpretative behavior – a logically 'would be' fact of response. The habit embodied in the object allows a semiotic system to interpret the sign as indicative of a class of entities or phenomena (Queiroz & El-Hani, 2006). Meaning and meaning change are conceived as a constraining factor of possible patterns of interpretative behavior, through habit and change of habit. The mediation of a sign results in a consistent relationship between variations in the form of the object and the corresponding effects on the interpreter (Queiroz et al., 2008).

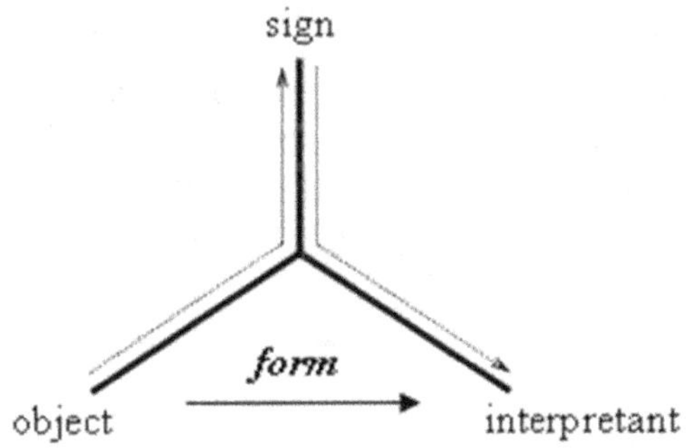

Figure 14.1. Semiosis as a relation between three irreducibly connected terms (sign-object-interpretant, S-O-I). This triadic relationship communicates a form from the object to the interpretant through the sign (symbolized by the horizontal arrow). The other two arrows indicate that the form is conveyed from the object to the interpretant through a determination of the sign by the object, and a determination of the interpretant by the sign.

Sign-mediated processes show a remarkable variety. The construction of appropriate typologies of these processes is a requisite for a deeper and more refined understanding of cognition. In an attempt to advance in the understanding of semiotic processes, Peirce proposed several typologies, with different degrees of refinement and several relationships to one another. A basic typology in his framework differentiates between iconic, indexical, and symbolic processes.

Icons, indices, and symbols are differentiated by Peirce based on how the sign relates to its object, that might be defined as the item to which the interpretants are related by the mediation of the sign (see Savan, 1987). This typology exhibits a property capable of functioning as an operational criterion to distinguish different kinds of signs: the relative dependence of sign-object-interpretant (S-O-I) components in triadic relation (Queiroz, 2012a, 2012b).

A symbol is an S-O relationship logically dependent of I. This relation has been characterized as a law ascribing S to O. A symbol is "a Sign (q.v.) which is constituted a sign merely or mainly by the fact that it is used and understood as such, whether the habit is natural or conventional, and without regard to the motives which originally governed its selection" (CP 2.307). Differently, an index is dependent of O. The relation between S and O has been characterized as one of contiguity: constraints resulting from the space-time existence of the object – irrelevant in symbolic processes – are the reason for the representation of O through S. In that case, S is really determined by O, in such a way that both must exist as events. The notion of spatio-temporal co-variation is the most characteristic property of indexical processes. When S is an icon, S signifies by means of qualities of S. Icons are dependent on the material, form and structure that are made: "An Icon is a sign which refers to the Object that it denotes merely by virtue of characters of its own, and which it possesses, just the same, whether any such Object actually exists or not" (CP 2.247). This relation between S and O, based on the qualities of S, has been characterized as one of similarity. The problem with the notion of similarity, however, is that it is too vague (see Stjernfelt, 2007). In order to detrivialize the notion of icon as a sign based on similarity, it is possible to give an operational definition of the icon.

Type of Sign	S-O Relation	S-O-I Dependence	
Icon	Similarity	Monadic (S)	Dependent on intrinsic properties of S
Index	Contiguity	Dyadic (S-O)	Dependent on S-O spatio-temporal correlation
Symbol	Law	Triadic (S-O-I)	S-O dependent of I mediation

Table 14.1: The fundamental types of signs underlying meaning processes – icons, indexes, and symbols. They are characterized in terms of relative dependence of sign-object-interpretant (S-O-I) components in triadic relation. The icon is the sign whose relevant properties for signification are its own intrinsic qualities: S depends on S.

The icons' dependence on its own materiality makes them suitable for modeling and experimentation. When an operational criterion is adopted, the icon is defined as anything whose manipulation can reveal more information about its object. Algebra, syntax, graphs, and formalizations of all types should be recognized as icons. This definition is considered a detrivialization of the notion that the icon is fundamentally based on a relation of similarity (see Stjernfelt, 2007; also, Hookway, 2002).

The key of iconicity is not perceived resemblance between the sign and what it signifies, but rather the possibility of making new discoveries about the object of a sign by observing features of the sign itself. Thus, a mathematical model of a physical system is an iconic representation because its use provides new information about the physical system. This is the distinctive feature and value of iconic representations: a sign resembles its object if, and only if, a study of the sign can yield new information about the object (Hookway, 2002, p. 102).

The icon is notably related to situatedness and distributedness of reasoning. It is the sign whose signification is S-dependent (i.e., dependent on the sign itself) and allows, through its manipulation, some discovery about the object. The notion of iconicity attests the capacity of material features to be the semiotic basis of cognitive operation, and not only play a secondary role.

Inferences are also understood as semiotic processes and have a place reserved under Peirce's typology. They are classified into three irreducible types – abduction, deduction and induction – corresponding to three subsequent phases in the process of scientific inquiry (CP 6.469-473). Abduction rises from the observation of a mass of facts that doesn't fit into the habits and expectations of the observer and culminates with the formation and selection of a hypothesis. Deduction develops testable consequences of the previously generated hypothesis. Based on these consequences, induction performs tests to evaluate it.

The characterization of abduction as the transformation of a mass of facts into hypotheses and the first stage of inquiry brings it close to perception (see Hoffmann, 2000; Paavola, 2005). For Peirce, perception involves an interpretative process (CP 5.181). It is through an inferential-like perceptual judgment that percepts are subsumed under general classes. This perceptual judgment accounts for the transformation of sense data into knowledge applicable to theoretical or practical use. It is subconscious, but if it was subjected to logical analysis, it would present an inferential – abductive – form (CP 5.181). Therefore, "all that makes knowledge applicable comes to us via abduction" (MS 692).

As an "act of insight" that "comes to us like a flash" (CP 5.181), abduction is germane to creativity. For Peirce, abduction is also the logical inference by which new knowledge can be obtained: "Abduction consists in studying the facts and devising a theory to explain them. It is the only kind of argument which starts a new idea" (CP 2.96). Magnani (2005) introduces the concept of "manipulative abduction" to refer to those cases where the inference depends on the exploration of external resources – it "happens when we are thinking through doing and not only, in a pragmatic sense, about doing" (Magnani, 2005, p. 274). According to Paavola (2011), in abduction the iconic character of reasoning is more prominent. Icons, abductions and perceptual judgments all have important similarities between themselves.

In all of them, some characteristics or phenomena suggest a potential way of interpreting or explaining these characteristics or phenomena and bringing them into some kind of an order (Paavola, 2011, p. 305).

Paavola has referred to these characteristics that only suggest a way in which they could be interpreted as "clue-like characteristics". In abduction, these clue-like characteristics, together with background knowledge, lead to the conclusion of a hypothesis (i.e., a promising way of arranging a mass of facts). This is a distributed process whenever these clue-like characteristics are predominantly material qualities of external signs. Abduction is especially near to the conceptualization of distributedness because it is an inference which relies on a mass of percepted data for its conclusion.

Peirce has developed an externalist semiotic perspective of cognition, where mind is the result of manipulation of signs, and (i) manipulation is described by irreducible forms of inferences, and (ii) signs are classified by different morphologies. Abduction and iconicity correspond respectively to the categories of inference and sign processes in which the situated aspect of Peirce's conception of mind is especially conceptualized. Abduction is a weak form of inference (see Paavola, 2004) related to perceptual features, while the icon is the S-dependent semiotic process. This treatment suggests that a reconsideration of the embodied-situated paradigm's own philosophical foundations can behave in semiotic terms.

Peircean semiotic theory of mind neither restricts representations to symbolic semiosis and inferential processes to deduction and induction, as in orthodox representationalism, nor rejects representations and inferences as in anti-representationalism (see Table 20.2).

	Representa-tionalism	Anti-representa-tionalism	Semiotic Theory of Mind
Signs	Symbolic	No	Not only symbolic but indexical and iconic
Inferences	Deductive, inductive	No	Deductive and inductive and abductive
Locus	Internal	External	Inference relies on internal and external resources

Table 20.2: Comparison between orthodox representationalism, anti-representationalism and the Semiotic Theory of Mind.

3. How Peirce's concept of information relates to cognition?

Already in his early studies, beginning in the 1860's, Peirce explained cognition as based on his triadic model of logic. These initial attempts at defining cognition emphasize the role of general representations, or symbols, as the carriers of information within the same mind, as when someone is thinking about a concept, or among different minds, during the process of communication of the meaning of a concept. In both cases, cognition is the result of the synthesis of feelings (firstness), efforts (secondness) and notions (thirdness) that cooperatively allow for the growth of symbols (CP 7.580).

In fact, while equating cognition and symbolic information, Peirce explains that both are the result of mental inferential processes, certainly synthetic and ampliative in their nature, carried out by either abductions or inductions.

As explained in section 2, these inferences begin from the perceptual judgments that bring to the mind the first amounts of meaningful information. "Perception is the possibility of acquiring information, of meaning more", explains Peirce (CP 7.587). At any moment, a particular mind – or a community of minds, if they are engaged in communication – must have a state of knowledge about the reality into which they are immersed. This state of knowledge encompasses the extensions and comprehensions of all symbols that participate and compose a particular mind or group in communion of grounds and interests.

Peirce's concept of information is then taken to be a third logical quantity, besides the traditional extension and comprehension, which in traditional logic always bear a relation of inverse proportionality. If the quantity of extension denotes the known objects of cognition and the quantity of comprehension enumerates its known predicates, information is acquisition of new knowledge by which concepts are interpreted in novel ways. Interpretation is necessary for information to be produced.

Moreover, cognition is also closely related to learning and helps us understand how our minds can internalize new forms and predicates (implying the growth of comprehension), or apply already known forms and predicates to new objects of attention (implying the growth of extension). In any of the two cases, information is produced because the growth of either comprehension or extension does not imply in the diminishing of the other quantity. The law of inverse proportionality does not hold when we learn something new about the Universe.

Peirce early studies of cognition were still influenced by Kant's transcendental philosophy, although Peirce is adamant in denying the ultimate object of cognition as being an incognizable "thing-in-itself", preferring to adopt the view of a continuous and infinite process coming from the vaguest, but developing towards the most definite, while never actually having either a beginning nor an end (CP 5. 314). The cognitive "train of thought", be it within the same mind or among different minds engaged in conversation, is better characterized as symbolic semiosis and leaves little space for the role of icons and indexes. This symbolistic account of cognition will be reviewed as Peirce discovers the need of quantification (indexical denotation) and diagrams (iconic connotation) for symbols to function as carriers of information from the object to the interpretant.

While Peirce maintains this basic mentalistic structure of cognition throughout his career, his late theory of mind is much less psychological, although psychistic. His definition is that whenever genuine semiosis is in action, mind is necessarily involved. It suffices that some kind of habit governs the process, which implies the presence of final causes. Since the laws of nature are all habitual, Peirce is ready to admit that mind is pervasive in the universe. In the late decade of 1900, his definition of sign and semiosis is further generalized to cope with Peirce's realistic view of possibilities and potentialities. Peirce in 1909 then defined the sign as a "cognizable" that allows for information not only in actual but also in possible minds:

> *A sign is a Cognizable that, on the one hand, is so determined [...], by something other than itself, called its object [...], while, on the other hand, it so determines some actual or potential Mind, the determination whereof I term the Interpretant created by the sign, that that interpreting mind is therein determined mediately by the Object.*
> (EP II: 492)

Bear with me: representation is not a faculty of individual minds or of communities anymore, but a logical process than can be explained by the general laws of semiotics. A sign does not have to necessarily be a mental symbol embodied in living beings, much less connected to human minds. It suffices that a sign be a "cognizable" even if never actually cognized. So a fossil fish can be a sign of a geological phenomenon even if there were no living creatures in this planet to ever interpret it. The important condition is that it would be interpreted as so if a sufficiently intelligent mind would ever unbury it to bring an actual interpretant about. In this new scenario, information is analogous to the increase of complexity in any part the Universe, or the internalization of new characters by a species via evolution (CP 2.420). Cognition is then redefined to incorporate extreme realism and full modality.

An analogical machine such as a weather-cock is certainly a cognizable capable of transmitting information about the wind, once a sufficiently intelligent mind cognizes where it is pointing to. Even if the weather-cock pointing to some direction (say, North) is primarily composed by indexes and icons, it is certainly a cognizable proposition. This is proved by the fact that once a mind that understands the natural laws that govern the movements of winds looks at the pointing weather-cock, it will be able to gather the information about the direction of the wind at that moment.

> *The reference of a sign to its object is brought into special prominence in a kind of sign whose fitness to be a sign is due to its being in a real reactive relation, –generally, a physical and dynamical relation, – with the object. Such a sign I term an index. As an example, take a weather-cock. This is a sign of the wind because the wind actively moves it. It faces in the very direction from which the wind blows. In so far as it does that, it involves an icon. The wind forces it to be an icon [...]. It is in this way that these indices convey information. They are propositions. That is they separately indicate their objects; the weather-cock because it turns with the wind and is known by its interpretant to do so; [...] If the weathercock sticks and fails to turn, [it] will be false. But if this is known to be the case, they sink at once to mere icons, at best. It is not essential to an index that it should thus involve an icon. Only, if it does not, it will convey no information.* (MS 7, pp. 17-18)

Why is the weather-cock informative?

- An icon embodies the information (the movement and pointing of the arrow resembles the very direction the wind is blowing). This works as the depth (connotation, predicate) of the assertion.
- An index physically connects the blow of the wind and the pointing of the arrow of the weather-cock. This works as the breadth (denotation, subject or object) of the assertion.
- A symbol gives the influx of the habitual rule of causation between the pointing arrow and the direction of the wind. In logical terms, the symbol connotes truly what it truly denotes. The symbol works as the habitual law that governs the functioning of the apparatus, transmitting the form of the direction of the wind to a possible mind. Its interpretation is information.

Cognition is indeed the actualization of what is cognizable, as much as information is the actualization of what is potentially informative, through actual interpretation. To wit, a mere gaze at the weather-cock for some living creature that does not have a previous knowledge about how it works[3] will not produce actual cognition and no information will be actually transmitted from the weather-cock to an interpreter, but that does not rule out the logical essence of the semiotic process.

But what if a computational machine performs the actualization of what is cognizable by means of a program that provides the necessary previous knowledge about the represented object, transforming the potential information into an actual effect? That would imply in genuine cognition and real information being produced by the operation of such machine. Peirce's semiotic realism united to his modal take on cognition provides a logical and philosophical framework for the understanding of cognition as a general and pervasive feature of our intelligible Universe. Situatedness and embodiment might be the necessary conditions for the actualization of cognitions, but semiosis is the sufficient condition for its reality.[4]

---

[3] What Peirce calls collateral experience about the object of the sign (CP 8.179).

[4] Section 1 was written by Ivo Assad Ibri, section 2 by João M. Queiroz, and section 3 by Vinicius Romanini.

Bibliography

Bergman, M. (2000). Reflections on the role of the communicative sign in semeiotic. *Transactions of the Charles S. Peirce Society*, 36, 225-254.

De Tienne, A. (2003). Learning qua semiosis. *Semiotics, Evolution, Energy and Development*, 3, 37-53.

Hausman, C.R. (1993). *Charles S. Peirce's evolutionary philosophy*. Cambridge (GB): Cambridge University Press.

Hoffmann, M. (2000). Problems with Peirce's concept of abduction. *Foundations of Science*, 4, 271-305.

Hookway, C. (2002). *Truth, rationality, and pragmatism: themes from Peirce*. Oxford University Press.

Ibri, I.A. (2010). Peircean seeds for a philosophy of art. In: Pelkey, J.; Haworth, K.A.; Hogue, J. & Sbrocchi, L.G. (eds.). *Semiotics 2010 "The Semiotics of Space"*. New York: Legas, pp. 1-16.

——— (2011). Semiotics and epistemology: the pragmatic ground of communication. In: Calcaterra, R.M. (ed.). *New perspectives on pragmatism and analytic philosophy*. Amsterdam: Rodopi, pp. 71-82.

——— (2012). Choices, dogmatisms and bets: justifying Peirce's realism. *Veritas*, 57(2), 51-61.

——— (2015). *Kósmos Noetós: a arquitetura metafísica de Charles S. Peirce*. São Paulo: Paulus.

Magnani, L. (2005). An abductive theory of scientific reasoning. *Semiotica*, 153, 261-286.

Paavola, S. (2004). Abduction through grammar, critic, and methodeutic. *Transactions of the Charles S. Peirce Society*, 40, 245-270.

——— (2005). Peircean abduction: instinct or inference? *Semiotica*, 153, 131-154.

——— (2011). Diagrams, iconicity, and abductive discovery. *Semiotica*, 186, 297-314.

Peirce, C.S. (1931-35, 1958). *Collected papers of Charles Sanders Peirce*. 8 vols. Ed. C. Hartshorne, P. Weiss & A.W. Burks. Cambridge (MA): Harvard University Press. [Cited as CP, followed by volume and paragraph number].

——— (1998). *The essential Peirce: selected philosophical writings*. The Peirce Edition Project (Ed.). Bloomington: Indiana University Press. [Cited as EP, followed by volume and page number].

——— (1982-2010). *Writings of Charles Sanders Peirce: a chronological edition*. The Peirce Edition Project (Ed.). Bloomington: Indiana University Press. [Cited as W, followed by volume and page number].

Queiroz, J. & El-Hani, C. (2006). Semiosis as an emergent process. *Transactions of the C.S. Peirce Society*, 42(1), 78-116

Queiroz, J.; Emmeche, C. & El-Hani, C. (2008). A Peircean approach to information and it's relationship with Bateson's and Jablonka's ideas. *American Journal of Semiotics*, 24, 75-94.

Queiroz, J. (2012a). Complexification. In: Favareau, D.; Cobley, P. & Kull, K. (eds.) *A more developed sign: interpreting the work of Jesper Hoffmeyer*. Tartu: Tartu University Press, pp. 67-70

——— (2012b). Dicent symbols in non-human semiotic processes. *Biosemiotics*, 5, 1-11.

Romanini, V. (2014). Semeiosis as a living process. In: Romanini, V. & Fernandez, E. (eds.). *Peirce and biosemiotics: a guess at the riddle of life.* New York: Springer, pp. 215-239.

Stjernfelt, F. (2007). *Diagrammatology: an investigation on the borderlines of phenomenology, ontology, and semiotics.* Synthese Library 336. Dordrecht: Springer.

Chapter 15

A dialogue concerning the mind-body problem

Alfredo Pereira Jr.
Department of Education, Institute of Biosciences
São Paulo State University (Unesp), Botucatu, SP, Brazil
apj@ibb.unesp.br

Jonas Gonçalves Coelho
School of Architecture, Arts and Communication
São Paulo State University (Unesp), Bauru, SP, Brazil
jonasgcoelho@gmail.com

Osvaldo Pessoa Jr.
Department of Philosophy, FFLCH
University of São Paulo (USP), São Paulo, SP, Brazil
opessoa@usp.br

Abstract

Three monist views on the mind-body problem are presented. Triple-aspect monism considers that there are three main phases of actualization of the potentialities in Nature: the physical, the informational, and the conscious. The double-face view assumes that conscious mind and brain are irreducible to each other, stressing not only that the conscious mind is dependent on the brain, but that changes in the brain are also dependent on the conscious mind. Qualitative physicalism adopts the mind-brain identity thesis, and defends the view that subjective qualia are actual physical attributes of some region of the brain.

1. Introduction

Osvaldo: Let us enjoy the boat trip from Ilha Grande (Big Island) to Angra dos Reis, in the coast of Rio de Janeiro, chatting about *the place of mind in the physical world.* I would like to start by asking whether we all agree that there was no mind in Nature, say 4 billion years ago, before the emergence of life on Earth.

Jonas: Yes. Roughly speaking, evolutionary biology allows us to state that the emergence of the conscious mind is relatively recent in the history of the Universe, only getting into existence when and where matter acquired, by genetic inheritance and natural selection, a particular kind of organization, a nervous system with a specific level of complexity. Based on this approach, it is reasonable to defend that the conscious mind is not a fundamental entity, independent and separate from the physical world, but a specific phenotype, i.e. an adaptive property of living matter, through which it has been acquiring, in its interaction with other physical and biological beings, specific levels of complexity.

2. Triple-aspect monism: an overview

Alfredo: I would like to speak from the perspective of *triple-aspect monism* (TAM), which intends to overcome the dichotomy of between materialism and idealism, as well as dualist attempts to save the appearances of both (see Pereira Jr., 2014). The ontology of TAM is based on the three main phases of actualization of the potentialities of Nature. The potential states are actualized in three ontological categories, according to an order (each aspect being a condition of possibility for each other). TAM does not attempt to explain the three categories in a metaphysical fashion; they are assumed to be fundamental aspects of reality, being actualized in any place and time when adequate "conditions of possibility" are satisfied.

TAM focuses on scientific concepts that afford explanations for the conditions of possibility of each aspect: a) Physical: refers to matter/energy in space/time. b) Informational: refers to complex forms being transmitted between physical systems. c) Conscious: refers to experiences of systems in which there is a feeling about the content of the information they process.

The immediate actualization of elementary energy forms is the physical aspect, characterized by the presence of matter. The structure of the physical space-time is defined according to the structure of matter/energy. In the universe or multiverse, there are regions where only the first aspect is actual. For instance, in a planet where material systems are in thermodynamic equilibrium, evolution stops in the first phase. In other planets there may be low entropy, and then the second aspect becomes actual (for instance, chemical clocks spontaneously appear).

Osvaldo: Okay, so we all agree that at a certain time, on planet Earth, there were no minds, and somehow they emerged. Your triple-aspect monism seems to me to be an original position. It is a form of monism, to which I agree, since I spouse a radical form of materialism, called "qualitative physicalism".

3. The double face view: an introduction

Jonas: My view is also monistic, in the sense that there is only one "substance", but is a form of Property Dualism. (I) Assuming that consciousness is the essential property of the mind, I propose a "double face view" of the mind-brain relationship, according to which *conscious mind* and *brain*, besides being essentially *distinct and irreducible*, are *dependent and inseparable*.

Before presenting this view, let me specify the notion of "mind" here considered. I assume both that consciousness is the essential property of the mind, the "central mental notion" (Searle, 1992, p. 83), and that conscious states always have contents: someone is conscious of something. So, we can ask "what is it like to be an individual *A*", paraphrasing Thomas Nagel famous article (1974), if we want to know how is the *subjective experience* or *phenomenal consciousness* of the individual *A*, including sensations, feelings and thoughts. For example, we may ask how is it to *feel* thirsty, sad or guilt, to *see* a red apple, to *pay attention to* a dangerous animal, to *remember* a scary experience, to *imagine* the future, to *reason* to win a chess game or to solve an equation, to *have concepts* such as color, time and black hole, to *believe* in God, to *wish* to get married, and so on.

Osvaldo: William Hamilton (1877, p. 191), the Scottish philosopher, claimed that one cannot actually "define" consciousness, since it is at the root of all knowledge. You are giving a kind of ostensive definition, saying that consciousness is that what you are experiencing right now!

Jonas: Yes. (II) But let me talk very quickly about the first pair mentioned before, the *distinction and irreducibility* between conscious mind and brain, a thesis assumed by the "double face view". I think that it is legitimate to say that conscious states and processes are not physical, in so far as they are not described by properties usually ascribed by physics to the physical world, like waves, particles, forces, etc. It is known through introspection that conscious states and processes exist and, at least in some degree, it is known what they are, even being difficult to describe them, and one does not need any kind of scientific knowledge to understand what they basically are, qualitatively speaking; on the contrary, science will depend on the limited description derived from introspection to find out the physical correlates of the events and processes of the conscious mind. So, according to the "double face view", the conscious mind cannot be reduced to physical events like brain states or processes, by identification or elimination, and just be explained as a property of brain states and processes in their interactions with the body and the physical and sociocultural environment. However, as I will argue ahead, this is not a Cartesian view, that is, *conscious mind* and *brain* are not two substances, which, although connected, could exist separately and independently of each other.

4. Qualitative physicalism

Osvaldo: I disagree that mind and brain cannot be considered identical. But what the heck, every philosopher of mind has his own metaphysical position! Here we are, on a small motor boat sailing through rather rough waters in the Big Island Bay, three philosophers of mind returning from a conference in some small corner of the Earth, as thousands of others do all around the world, each with a different opinion. Will there be some scientific revolution in Neuroscience, Psychology and the Philosophy of Mind, that will unify our views on the mind/body problem?

I believe so. And I think the beginning of the revolution will take place when empirical scientific research establishes where in the brain is the seat of consciousness, what part of the brain is the immediate (or most proximal) cause of consciousness, or identical with it, as in my view. My view is also monistic, but I attribute a special role to the *mind-brain identity thesis*. When we see a red pitanga berry, as those we avidly ate at Two Rivers Beach, the redness is not in the actual fruit, but somewhere in our mind, as formulated by Galileo and subsequent classical philosophers. This is an "internalist" view on colors and other qualia. But assuming the mind-brain identity thesis, it follows that colors are located in our physical brains. Adopting physicalism, one finally concludes that colors and other sense data and emotions are real properties of our brain tissue.

Jonas: If you say that redness is in your brain, why doesn't your brain tissue actually look red? I mean, forgetting that it's full of blood.

Osvaldo: When you look at a pitanga in sunlight, light is absorbed selectively on the surface of the berry, with longer wavelengths of light being reflected to your eye. However, when you look at the region where redness is produced in someone's brain, the tissue does not selectively reflect light in the same way as the pitanga does, so it will not look red for an outsider. This metaphysical position, called the "colored-brain thesis" by Stubenberg (1998, p. 169), actually put forth by Thomas Case (1888), and which I rather name "qualitative physicalism", claims that subjective qualia or sense-data are physically as they appear. This is a materialist version of "Russellian monism" (Alter & Nagasawa, 2012), as developed by Mach, Schlick and Russell: science has only access to the structure of reality, to the relations between things, while our only access to the thing-in-itself or to the inscrutables of reality is in our subjective experience, which these authors take to be sense-data, in the so-called "neutral monism".

Inspired by Mary's room thought experiment, qualitative physicalism separates the quantitative-linguistic description of colors that Mary has inside her room devoid of colors (a theoretical scientific description) from the qualitative experience she has when she actually sees a color for the first time (acquaintance). There is clearly an explanatory gap between the qualitative experience and the theoretical description, which neuroscience in the future will bridge with psychophysiological bridge laws that will have to be taken as principles (therefore unexplained).

5. The three actualizations, according to TAM

Alfredo: I follow Jonas in distrusting the identity of mind and body. But I think one needs to include a third aspect to account for reality, which is information. In triple-aspect monism (TAM), Nature is conceived as the totality of existence; a mind-independent eternal reality, composed of interacting, self-organizing elementary energy forms. The elementary forms can be compared to an "alphabet" of reality – or to Plato's Ideas, in the Aristotelian interpretation (they are not placed in a separate realm, but embodied in Nature). Elementary energy forms are claimed to exist eternally *in potential states* (another difference from Plato's Ideas). We do not observe or experience the elementary forms themselves, but only their combinations. Their practically infinite combinations define different kinds of evolutionary processes that occur in different regions (for instance, in the planets of the solar system).

The ensemble of all possible combinations of elementary waveforms define the State Space of Nature, a N-dimensional structure where all possible dynamical evolutions of Nature can be identified. In TAM, the quantum wavefunction is conceived as composed of real, interacting, superposed and entangled elementary energy forms. At each measurement, the complex of interacting forms decoheres, generating a macrostate with fixed properties (quantitative and qualitative). We can directly measure and register the properties of the resulting macrostates.

"Self-organization" refers to the activity of systems composed of dynamical interacting independent units generating new phenomena. Cournot's concept of *chance* was used by Debrun (1996) to approach the dynamics of self-organizing systems. It is not identical to the concept of *randomness*, but refers to the semi-deterministic process of interaction of statistically independent units, generating what could be called "relational collectives" (Lungarzo & Pereira Jr., 2009).

Osvaldo: So, this is the first aspect, the physical one, that existed before living creatures. I assume that when you talk about "quantum measurement", you also include "environmental monitoring", which also leads to decoherence without the presence of an intelligent being.

Alfredo: Yes. Now the second actualization builds on complex forms generated by the combination of the elementary ones. This aspect of Nature contains all laws and principles of Nature: mathematical relations, logical inferential rules, information patterns, etc. Rules of Nature do not exist in the void; they are the results of the systematic self-organization of the elements. This aspect was largely expanded with the emergence of biological systems that exchange all kinds of signals internally, with other living systems and with the physical environment.

The third actualization we know about is conscious experience. It occurs when a system feels the information it processes. When complex patterns are formed and transmitted between material systems, the second aspect is present, but it is not sufficient for the third aspect; something more is needed: the existence of systems able to feel the meaning (for themselves) of the informational messages (e.g., to recognize regularities in the chemical clock as temporal rhythm). Feeling is an affective state characterized by the capacity of the information to *globally* affect (or to *cause*, in the sense of Aristotle's Formal Cause) the physical structure of the system.

Osvaldo: I see... It is curious, because I would consider the informational aspect as simply the patterns or arrangements of matter, something pertaining to the material realm.

6. The double face view: inseparability of mind and brain

Jonas: Well, let me continue presenting the "double face view" of the mind-body relationship. I have stressed the distinction and irreducibility between mind and body. (III) Now I will talk about the *dependency and inseparability* between *conscious mind* and *brain*, the main point I wish to stress here. First, to explain the *conscious mind* it is necessary to consider the *brain* (conscious mind → brain), for the structural and functional organization of the *brain* is the cause not just of the existence, but also of the properties – form and content – of the conscious mind.

(IV) Second, to explain the structure and functioning of the *brain* the *conscious mind* has to be considered (brain → conscious mind), for the *conscious mind* is responsible, at least in part, for the structural and functional organization of the brain. Let me develop little bit more this "double face view".

(V) The thesis that the structural and functional organization of the brain is the *immediate cause*[1] of existence, form and content of the conscious mind –

[1] To consider the brain as the immediate cause of the conscious mind does not mean to disregard its mediate causes, such as the body and the physical and sociocultural environment, without which the brain would not exist and would not be structurally and functionally what it is.

conscious mind as brain – has been endorsed by evolutionary biology and neuroscientific research.

The assumption that the brain is the immediate cause of the conscious mind has also been supported by countless neuroscientific studies, most of them published in the last 30 years, which use different methods of research and a variety of technologies that allow to peer into the normal and abnormal brains during and after cognitive, affective and behavioral tasks. This sort of research has fostered the development of therapeutic practices targeting cognitive, affective and behavioral dysfunctions, by manipulating the brain, for example, chemically and electromagnetically, changing its structure and functioning on both levels, macrophysical and microphysical. Pointing to the same direction, there is a wide research program, which includes: studies of brain injuries and its implications to emotional, cognitive and moral behavior; brain-machine interface; cranial electromagnetic stimulation, and the development of drugs targeting mental illness, emotional states, cognitive skills and moral behavior.

Osvaldo: I subscribe to this view. We could say that the mind "supervenes" on the brain (and not only on the body or the whole environment). One could push this even further, and postulate a localizationist thesis: that phenomenal consciousness also supervenes on a rather small region of the brain. But this issue is an open question for neuroscience, whether the proximal cause of consciousness is localized (Bogen, 1995, suggested the thalamus, others the posterior prefrontal cortex) or whether it is distributed as a dialogue between thalamus and cortex (Llinás & Ribary, 2001).

Oh, I'm feeling seasick, with the rolling waves.

Jonas: Lie down on the floor, belly up! Let me continue.

(VI) So, I think that there is sufficient evidence to assume that the conscious mind is a property of the brain, and as I will argue next, an embodied and situated physically and socioculturally brain, so to explain how the brain produces the conscious mind would depend on understanding how the brain works, no doubt a tough problem, but not an insoluble mystery.

The other thesis of the "double face view", that the structural and functional organization of the brain depends on and is inseparable from the conscious mind – *brain as conscious mind* – is also supported by contemporary research in neuroscience, from which it can be inferred that the brain, besides being physically embodied and situated, is also conscientiously embodied and situated. I mean that, via conscious mind, though not exclusively, the brain interacts with its body and with the physical and the sociocultural environment in which it is immersed. The sense organs and proprioceptive/interoceptive structures of the body are here considered as a kind of window that allows the brain, which is inside of the head, to interact with the body, where it is located, and with the physical and sociocultural environment, where it is situated, being consciousness

– the essential property of the mind – one specific property of the brain, which is present in all of those windows. It is known that the brain interacts unconsciously with physical stimuli through the sense organs and proprioceptive/interoceptive mechanisms, however I would like to stress here the ability of the brain to interact via consciousness with both, the physical and the sociocultural environment. What the brain is and what the brain does is inseparable from its structural and functional architecture, genetically constructed and, as studies of brain plasticity show, resultant of brain interactions with the physical and sociocultural environment.

So, the structure and functionality of the brain depends on both, preinstalled cognitive and emotional capabilities built genetically, and the physical/sociocultural environment, which affects the brain thanks to its ability to interact with the body and with the external world via its conscious mind. Regarding this aspect, contemporary neuroscience has shown that the brain is not only modified when it is physically altered by sensory stimuli, injury, surgery, chemical and electromagnetic manipulation, etc., but also, thanks to its plasticity, by many different kinds of physical and psychological experiences. As shown by a paradigmatic definition of brain plasticity, there is an essential relationship between brain changes and experience: "we can define brain plasticity as the capacity of neurons and of neural circuits in the brain to change, structurally and functionally, in response to experience" (Sale et al., 2014, p. 190). What neuroscientists in general do not specify, but I am considering as fundamental, is that what is said about *experience* can be interpreted, at least in many cases, as conscious experience. So, I am here specially thinking about the relationship between *conscious experiences* and brain plasticity. By exposure to sociocultural practices, such as those related to a regular education, a child, by mean of its conscious mind, has its brain modified, so the sociocultural learning is not an abstraction, separated from what happens in the brain.

7. The material correlates of consciousness

Alfredo: The identification of sufficient conditions for consciousness is a huge task for scientific research. According to TAM, feelings are related to wavelike processes that affect living systems. Plant tissue, having the morphology of a syncytium, may probably be able of forming hydro-ionic waves that would correspond to a primitive sensibility to conditions as environmental heat or sound. In the brain and body of animals, sentience is possibly related to wavelike activities in glial cells, extracellular and cerebral fluids, blood and muscles. The evolution of neuronal networks endows animals with more and more powerful cognitive capabilities – instantiated in specialized neuronal circuits – that expand the capacity of feeling that begun with plants. In the human species, consciousness can be conceptualized as *cognition with feeling* (or feeling improved by cognition).

There are two kinds of information processing in the brain. One is by means of electric pulses (action potentials) in neuronal networks. The other is by means of hydro-ionic waves guided by proteins, in glial cells, extracellular medium, cerebrospinal fluid and blood flow. While the main physiological activity of neurons is the electrical action potential – composed of fluxes of sodium and potassium ions through the membrane, activating molecular and biochemical processes at the synaptic cleft –, astrocytes have internal calcium waves that cross the entire network and control the concentration of extracellular potassium.

Neurons receive and process information from the environment and the body, and control behavior, while astrocytes receive information from neurons and blood, but do not directly control behavior. There is an increasing evidence for the hypothesis that brain tissue waves, having the astroglial network as a "Master Hub" (Pereira Jr. & Furlan, 2010), instantiate conscious feelings and modulate neuronal cognitive processes, according to the valence of the feeling (reinforcing when it feels good, and depressing when it feels bad). The above hypothesis illustrates how conscious activity is instantiated in the living body according to TAM. The feeling of qualia (e.g., color) comes from the dynamic waveform instantiated in brain tissue, having the astroglial network as the hub that mediates the effect.

The philosophical concept of the relation of the conscious mind with brain activity that results from TAM is that conscious activity is a fundamental aspect of reality, not (epistemologically or ontologically) reducible to the physical or informational aspect, but requiring the actualization of both (i.e., requiring the existence of physical systems that process information and feel the meaning attributed to the message).

8. The mind-brain identity thesis

Osvaldo: I have a similar view concerning what you called "instantiation of conscious feeling" in the brain tissue. Relying on the mind-brain identity thesis, I would say the redness of the pitanga berry is identical to a certain region of the brain, which consists of tissue made of nerve cells (neurons and glia) fed with complicated spatiotemporal patterns of electrochemical impulses from the neurons. The patterns themselves, which could be generated in a computer, are not sufficient to emulate consciousness, since biological matter is taken to be essential for consciousness. This is a version of Searle's (1992) "biological naturalism", a position that is opposed to "machine functionalism".

The mind-brain identity thesis seems necessary for any materialist view. If a thought arises from the brain but is not identical to a part of it, then it must be hovering above the brain tissue, but where is it located in the physical realm? Maybe it is an electromagnetic field produced by the brain, but then it is identical to the field. Maybe it is an emergent property of the brain tissue, but then one would have to admit that there are also emergent physical properties, so the simplest explanation for the physicalist would be that the mind must be identical to such an emergent physical property. The identity thesis seems inescapable for the materialist, but it should be used appropriately, and not as done by Paul M. Churchland (2007, p. 120), who proposes that "the sourness of a spoonful of lemon juice [...] is identical with the relative concentration of hydrogen ions in that liquid". According to qualitative physicalism, the sourness is identical to the physical components of a certain internal brain state, not to the external physical state of the juice or of the tongue.

Churchland's view is similar to U.T. Place's, in attributing to subjective qualia a secondary status, something that is clearer in Feyerabend's (1963) version of identity theory, which eliminates mental concepts altogether. Place's identity thesis is in fact an eliminative thesis, as is clear in his statement that there is a "phenomenological fallacy" in the claim that the greenness of an after-image experienced by someone is "a literal property of objects", since there isn't "anything green in his brain" (Place, 1956, p. 49). The famous example that Place gives of the identity thesis, that "lightning is the motion of electric charges" (p. 47), is the comparison of two linguistic-quantitative statements, and is therefore non-problematic. The problem arises when comparing a subjective experience and a physical description of the brain. The identity thesis is the correct step, but Place was not willing to buy the colored-brain thesis with which Boring had flirted, "which for a philosopher would constitute an immediate knockdown reductio ad absurdum of his position" (Place, 2000, p. 1). The claim that a subjective state of pain is identical to a neurological state should not eliminate the subjective qualitative state; one should bite the bullet and recognize that the pain is a physical attribute of the brain (since the mental state is identical to the physical state).

There are many versions of the mind-body identity thesis. Any monistic view must identify mind and brain in a fundamental level. One may claim Spinoza had an identity view, since for him, as well as for Fechner and more recently Thomas Nagel, mind and brain are two aspects or perspectives of a single fundamental entity. The first expression of the identity thesis appeared with the materialist Democritus, and according to Aristotle (*De Anima*, I.2, 405a9) "soul he regards as identical with mind".

9. The double face view and the causal role of the conscious mind

Jonas: From what was said about the relationship between *conscious mind* and *brain*, the "double face view", I infer that the causal action of the first on the second is not unintelligible, and is not a mystery as well. What turns mental causation unintelligible is a misleading formulation of the problem, which does not distinguish *dualism of property* from *dualism of substance.* In a formulation of the problem of mental causation, "how can the conscious mind, being a non-physical property of the brain, act causally on the brain?", the conscious mind is treated as if it were a non-physical substance, which, by itself, would have a causal power on the physical brain, so being really an unintelligible relationship. However, taking seriously the idea that the conscious mind is a property of the brain, a distinct formulation of the problem of mental causation would be more appropriate, like: "how does the brain, guided by its conscious mind, act causally on the brain?". Now the conscious mind is considered as a non-physical property of the brain, by means of which the brain acts upon itself. What I am proposing here is to focus on the causal power of the brain derived from the fact that it has a conscious mind, which requires, as said before, an understanding of how the fact that someone has a conscious mind determines the structure and functionality of the brain and, consequently its intentional action on the body and external environment.

So, to accept that the brain plays an essential causal role, being causally responsible for conscious events and processes, does not imply the exclusion of the causal power of the conscious mind, that is, it does not render the conscious mind a mere epiphenomenon or "shadow" of the brain. Even if one accepts that the contents of consciousness depends on brain structure and functioning, in its relationship with its body and external physical and sociocultural environment, that does not prevent the conscious contents of being responsible for changes in the structure and functioning of the brain. To sum up, according to the double face view here presented, the *embodied and situated brain* is the organ of the *conscious mind,* and the *conscious mind* gives the *brain* causal powers which the latter would not have without its *conscious mind,* being as legitimate to ascribe a causal power to the *conscious mind* (not by itself as if it were a substance) as it is justifiable to attribute causal power to the physical and sociocultural environment, which, by changing the *brain through its conscious mind,* provides the most part of the contents of the conscious mind.

Osvaldo: Our trip is ending, we're arriving at what once was a beautiful coastal town, Angra dos Reis, but which is now full of shanties...

Bibliography

Alter, T. & Nagasawa, Y. (2012). What is Russellian monism? *Journal of Consciousness Studies*, 19, 67-95.

Aristotle (1907). *De Anima*. Transl. R.D. Hicks. Cambridge (GB): Cambridge University Press.

Bogen, J.E. (1995). On the neurophysiology of consciousness. I. An overview. II. Constraining the semantic problem. *Consciousness and Cognition*, 4, 52-62, 137-58.

Case, T. (1888). *Physical realism: being an analytical philosophy from the physical objects of science to the physical data of sense*. London: Longmans, Green & Co.

Churchland, P.M. (2007). On the reality (and diversity) of objective colors: how color-qualia space is a map of reflectance-profile space. *Philosophy of Science*, 74(2), 119-149.

Debrun, M. (1996). A ideia de auto-organização. In: Debrun, M.; Gonzalez, M.E.Q. & Pessoa Jr., O. (eds.). *Auto-organização: estudos interdisciplinares*. Campinas: Centro de Lógica e Epistemologia/UNICAMP, pp. 3-23.

Feyerabend, P.K. (1963). Comment: Mental events and the brain. *Journal of Philosophy*, 60(11), 295-296.

Hamilton, W. (1877) *Lectures on metaphysics and logic*. 6th ed., vol. I. Edinburgh: William Blackwood. (Lectures delivered in 1836-37.)

Llinás, R. & Ribary, U. (2001). Consciousness and the brain: the thalamocortical dialogue in health and disease. *Annals of the New York Academy of Sciences*, 929, 166-75.

Lungarzo, C. & Pereira Jr., A. (2009). A Cournotian approach to the emergence of relational collectives. *TECCOGS: Revista Digital de Tecnologias Cognitivas*, 1, 1-17.

Nagel, T. (1974). What is it like to be a bat? *Philosophical Review*, 83(4), 435-450.

Pereira Jr., A. (2014). Triple-aspect monism: physiological, mental unconscious and conscious aspects of brain activity. *Journal of Integrative Neuroscience*, 13(2), 201-227.

Pereira Jr., A. & Furlan, F.A. (2010). Astrocytes and human cognition: modeling information integration and modulation of neuronal activity. *Progress in Neurobiology*, 92, 405-420.

Place, U.T (1956). Is consciousness a brain process? *British Journal of Psychology 47*: 44-50.

——— (2000). Identity theories. Unfinished manuscript, edited by S. Schneider, available at the site *A field guide to the philosophy of mind*: http://host.uniroma3.it/progetti/kant/field/mbit.htm

Sale, A.; Berardi, N. & Maffei, L. (2014). Environment and brain plasticity: towards an endogenous pharmacotherapy. *Physiological Reviews*, 94(1), 189-234.

Searle, J. (1992). *The rediscovery of the mind*. Cambridge (MA): MIT Press.

Stubenberg, L. (1998). *Consciousness and qualia*. Amsterdam: John Benjamins.

Chapter 16

Machine learning and the probabilistic modeling of cognition and behavior

Peter Maurice Erna Claessens
Center of Mathematics, Computation and Cognition (CMCC)
Federal University of ABC, São Bernardo do Campo, SP, Brazil
peter.claessens@ufabc.edu.br

João Ricardo Sato
Center of Mathematics, Computation and Cognition (CMCC)
Federal University of ABC, São Bernardo do Campo, SP, Brazil
joao.sato@ufabc.edu.br

Abstract

This text is a discussion of machine learning, especially probabilistic modeling using Bayesian statistical inference, based on three general questions. The authors acknowledge the importance of Bayesian methods, but recommend that they be complemented by other approaches. Automated decision making in the medical sciences is also examined, as well as aspects of the machine techniques of Deep Learning, the training procedures of which are compared to the maturation of the human brain.

1. Mathematical modeling of cognition

In the project of modeling cognition using mathematics, among the many mathematical structures at our disposal, which should be used? Should Bayesian statistical inference be preferred?

Peter: In the first place, it is important to make a distinction between the use of mathematical constructs in a methodological sense, to help understand the cognitive architecture and neural implementation of cognitive processes, on the one hand, and the hypothesis that mathematical constructs are direct equivalents of elements of cognitive processing. One can, for example, use methods available in graph theory to study cortical connectivity without assuming that the results of analyses of this type reveal a fundamental link between the structure of cognition and graph-theoretical notions. Due to its power to reveal underlying patterns, support vector machines (SVM) are used and will continue to be used to investigate patterns of cortical processing, without any need to suppose that, at an algorithmic level, natural information processing amounts to the same as the kind of pattern recognition SVM excel at (e.g. Sato et al., 2012).

Bayesian statistical inference, however, has the peculiar characteristic of defining at the same time a family of methodological tools in the study of cognition – in the same way as it has been used in other fields of data-drive science that are completely unrelated to cognition – as well as a self-contained functional model of the underpinning of cognitive processes, from low-level sensation, over perception of time, to high-level reasoning. The reason why the Bayesian paradigm is this promising relates to Bayesian inference being the normative theory on reasoning under probabilistic uncertainty. At least in theory, this should mean that reasoning according to the Bayesian framework is reasoning according to 'the best possible' manner. This makes it an excellent working hypothesis. If evolution through natural selection, which is a widely accepted powerful scientific theory, is a phylogenetic process that allows for organisms to reach the best adaptation in terms of anatomy and even behavioral patterns through generations, even in a changing environment, one might conjecture that there is no reason to believe that cognitive evolution would follow different principles. If this is true, in the absence of another theory that would be equally powerful in its explanations and predictions, natural information processing should also have evolved to lead to the best adaptation possible in a changing and especially uncertain environment. Since, from a statistical point of view, Bayesian inference amounts to optimal use of information, it is a natural contender for an encompassing theory of cognition. It is therefore no surprise that the Bayesian approach has taken cognitive science by storm, first as a model of perception (e.g. Knill & Richards, 1996), and later as a model of higher-order cognitive processes (e.g. Chater & Oaksford, 2008).

This is not to say that Bayesian inference is without criticism. Some cognitive scientists position themselves firmly against what they consider an unwarranted shift of paradigm, almost as if it were a leap of faith (e.g. Bowers & Davis, 2012; Jones & Love, 2011). Many phenomena seem incompatible with a Bayesian inference interpretation (e.g. Brayanov & Smith, 2010). Even the simplest Bayesian model requires the choice of a prior and a likelihood function. Minimally complex models of behavior often involve interacting priors and likelihood functions. Although in some cases it is possible to measure the appropriate prior distribution (e.g., for the visual judgement of line orientation, see Coppola et al., 1998), it is more often than not an arbitrary choice, inspired by convenience, tradition, or even fit to obtained data. Cognitive models also need to be interfaced to behavioral performance, and also this involves non-trivial choices. All of this means that, in the day-to-day challenges that cognitive modelers face to create a specific model for a specific task, relatively arbitrary decisions are made, and even without necessarily leaving the framework of cognition-as-statistical-inference, the complexity of the model can indefinitely be increased in order to bring the predictions closer to the data. At one extreme of this set of choices, some critics will claim, Bayesian cognitive science is not falsifiable: failure to model behavioral data as Bayesian inference can be attributed to oversimplification and wrong model choices. The set of Bayesian models that one can produce for one specific cognitive process is therefore so large that it makes better sense to think of Bayesian inference as an umbrella framework rather than as a particular strategy that will guide to a specific model. However, this is probably not a very different situation from the state of models that were and are being developed in probably the largest paradigm before – though not superseded by – the Bayesian approach, namely the Parallel Distributed Processing framework (PDP) (McClelland & Rumelhart, 1986; Rumelhart & McClelland, 1986).

A natural extension of Bayesian inference, Bayesian Decision Theory or BDT, relates probabilistic judgments to an optimal sequence of action upon choice of a loss function. But what exactly is a loss function in the context of an experiment with human observers? Often, humans make judgments that in a first analysis are not rational under any choice of loss function (Wozny et al., 2010), apparently in line with a strategy known as probability matching, in which humans (and some other animals, for example honeybees, see Niv et al., 2002) tend to choose among options proportionally with their posterior odds rather than the most probable one. Patterns similar to probability matching might also result from exploration behavior or attempts to apply a detected (even if non-existent) pattern to the sequence of events to be judged (Feher da Silva & Baldo, 2012). Inasmuch as Bayesian Decision Theory relates cognitive processes and motor action, it is the natural candidate for normative models of 'agent' rather than 'cogent' systems, bringing us closer to an embodied view of cognition, although, of course, radical supporters of embodied cognitive science will not be willing to accept the existence of explicit inference through representation of interpretations and probabilities (ideas explored in Clark, 2013, and its comments, and in Dotov, 2014).

João: Concerning the question of whether Bayesian statistical inference is to be preferred, indeed, in recent years Bayesian statistics has been a key player regarding inferences on brain processes and cognitive modeling (Friston et al., 2016; George & Hawkins, 2009). One of the main reasons is because this framework facilitates statistical modeling and the convergence of algorithms, even in the case of high dimensional data, such as in functional magnetic resonance imaging (fMRI), electroencephalography (EEG), near-infrared spectroscopy (NIRS) and magnetoencephalography (MEG). The consideration of previous knowledge is one of the main advantages of this approach. However, in my opinion, we should not constrain neural data analyses solely to Bayesian statistics. Since Bayesian statistics usually requires the specification of prior and likelihood functions, it is often necessary to make assumptions or arbitrary choices regarding the hyperparameters or data distribution. There are some cases in which these assumptions may not hold or is unknown (e.g., brain diseases, aging, development, etc.), and methods with weak assumptions are possible complementary methods and alternatives (e.g., permutation methods; Winkler et al., 2016).

2. Decisions in medical sciences

What are the best approaches for decision making in the medical sciences?

Peter: Decision making in medical diagnostic inference and therapeutic intervention is a complex process, possibly subject to the largest number of simultaneous factors to observe among all types of human decision. Several machine learning approaches have been tested in the medical sciences (see, for example, the special issue of *Machine Learning*, edited by Wiens & Wallace, 2016, and the exceptional 6-volume series on the topic edited by Cleophas & Zwinderman, 2013-14). The Bayesian network paradigm is an attractive framework in this context (for example, see the Hepar II system, for the diagnosis of liver disorders, Wasyluk et al., 2001, and the work on tinnitus diagnosis, Jangholi, 2014); as a combination of graph and probabilistic theory, it is apt to deal with interacting variables on the one hand, and probabilistic uncertainty on the other. As artificial neural networks, they can be trained based on a set of expert judgments, but represent the probabilistic nature of information more directly. Bayes nets can be exploited to reveal which diagnostic tests will provide most information about the underlying cause of the disease process, through calculation of conditional mutual information. In principle, Bayesian networks give 'the best possible' answers if all relevant variables, dependencies and conditional probability tables were known. Of course, this would be a utopian scenario. We can only try to get as close as possible by relying on a large enough set of training cases and human judgements. Conditional probability tables and in many cases also the structure of the network are generally estimated in a data-driven fashion. The success of the Bayesian network will therefore depend on the quantity and quality of the input data, and even in the presence of a large body of reliable cases, parameter and structure learning in the network can be limited for purely computational reasons. Some very efficient algorithms for the particularly difficult case of structure learning are available, such as the application of Chow-Liu trees to Bayesian belief networks (Cheng et al., 2002), but these do not guarantee that the best possible graph is obtained.

For these and other reasons, it is still difficult to envisage a future in which medical doctors are substituted by machines. Human diagnostics, on the other hand, is subject to a host of different problems, among which so-called cognitive illusions. This goes for any kind of inference, including those done by experts from other fields, so it is not reasonable to assume that clinicians are free from biases when making medical decisions (Croskerry, 2003). As an example, there is the case of confirmation bias – the tendency to give greater weight to an observation that confirms an existing conviction and less to an observation that is not consistent with the current hypothesis. This in turn leads to overconfidence, an estimate of certainty, about the subjectively most likely diagnosis, that is higher than it deserves to be. In combination with the common heuristic of entertaining only one hypothesis at a time, which is only removed if evidence is strongly contradictory, this might lead to late diagnosis of life-threatening diseases, such as pneumonia, which are relatively often confounded with very common ones (Graber et al., 2005; Schiff et al., 2009; Singh et al., 2013). Possibly human diagnostic reasoning is prone to biases because clinicians tend to use a pattern-recognition-like strategy for diagnosis, similar to what has been called Type-I or System-I cognitive processes, and not effortful conscious inference, called Type-II reasoning in cognitive psychology (Croskerry et al., 2014; see Evans & Stanovich, 2013, for a more general review). Under the supposition that Type-I reasoning is more error-prone than Type-II processes, it might be possible, and has been tried, to encourage medical students to engage in effortful reasoning rather than automatic pattern classification, but it is as of yet unsure that this leads to improvement in the diagnosis reached (Norman et al., 2014). Note that the inevitable occurrence of errors in human judgment leads to an interesting challenge or even paradox: given that practically all machine learning approaches to medical diagnosis are supervised, that is, depend on training with possibly noisy human judgments that are treated as 'ground truth', is it possible to design algorithms that are *better* diagnosticians than human doctors?

At the current time, it will be hard to find someone who will be convinced to go talk to a computer rather than a doctor in order to establish a diagnosis on some illness. It is certainly too early to hand over clinical diagnosis to machine intelligence, however at the same time it is difficult to imagine that the statistical patterns a computer algorithm might detect in training data obtained from several experts in quantities that go beyond what any of these experts will deal with in an entire life time are useless for diagnosis. An intermediate solution is to design diagnostic support systems (DSS) that assist the clinician in reaching a diagnostic conclusion, not realistically with the intention to perform better than the human doctor, but rather as an ideally continuously learning digital assistant in calculating odds and proposing the most likely alternative hypotheses to whatever the user is proposing. The success of a DSS does not only depend on its diagnostic success rate, but also, and critically, on its usability. Medical experts will not be willing to dedicate their time to the input of patient data, so there is certainly an important role for patient data that are digitalized from the source on, preferably in a standardized way, so the database can span across different clinical centers, and for image recognition and possibly even speech recognition software. Several initiatives, in different countries, move into the direction of a centralized patient data base, often in order to monitor the use of public health services. The advantages of fully using 'big data' approaches to medical records are significant, but data access will have to be counterbalanced by an ethical and legal framework that secures anonymity and other basic rights to the patients. In any case, and this goes for human as well as machine learning, procedures need to be put in place to guarantee that feedback on the success or failure of a diagnosis flows back to the expert, or computational system, who first made the diagnosis.

A final challenge to Bayesian networks and other machine learning approaches is to turn diagnostic inference into true decisions, which, different from inferential conclusions, are characterized by benefits and risks, which are, more often than not, subjective. Patients might weigh the same benefits and risks differently. A human medical doctor can interpret patient's verbal expressions of fear, for example, or concerns about quality of life, or expected life span, and take this into account in proposing a course of therapeutic action, or even in deciding which exams and tests are essential and which are not, taking both diagnostic value and patient concern into account. As of yet, we have much better formal frameworks to determine probabilities than we have to quantify attributions of value, one important reason why the true 'robot doctor' is still situated in the realm of science fiction.

João: Automated support-to-diagnosis systems are one of the main outcomes of technological advances in the last decades. Indeed, it was shown that with the use of these systems, it was possible to improve the accuracy of medical diagnosis (Doi et al., 1999). Classification and machine learning methods have been more frequent in these systems. However, independently of how high the accuracy of these technologies is, the ones based on black-boxes which provide solely the pipeline "user input-computation-output" will not be convincing for the clinicians. In most cases, the expert doctors request to understand the reasoning behind the outputs of these methods in the single case (for one specific patient, and not in average or probability distributions). Many doctors would argue that they could find (bizarre) counter examples in which the automated system would fail (e.g., someone with a rare disease or many confounders), if they can understand the algorithm behind the black-box. Thus, in my opinion, the best approaches for decision making in medical sciences are the ones that could provide not only high sensitivity and specificity, but also that are capable of providing a reasoning for the inductions made (e.g., decision trees). Also, systems that could improve the clinicians reasoning in the single patient case, or methods that might provide insights into the underlying physiopathological processes, are of great value. There is an argument that computer-aided diagnosis could provide a second opinion. This is true, but if it cannot explain why or how the decision was made, I do not believe clinicians would be convinced by an unexplained decision.

3. Deep learning

How promising is Deep Learning, in its relationship with neural computation? How is machine learning essential for discoveries in the maturation of young human brains, and conversely, how can findings in the maturation of young human brains affect the techniques in machine learning?

João: One of the most interesting properties inherent to many machine learning methods is that they generate predictions based on a (learned or trained) model. And model based prediction is a critical issue in the scientific method. Recently, it was demonstrated that by using fMRI data, it was possible to predict the subject's chronological age based on brain activation patterns (Dosenbach et al., 2010). Interestingly, by using these methods, it might be possible to identify and quantify deviations from the typical neurodevelopmental processes by building maturity indexes (Sato et al., 2015). Thus, for a child, if the chronological age is considerably higher than the one predicted by a machine learning method, this may suggest that the brain system (used as input) is somehow delayed, from a maturational perspective. Since many psychiatric disorders have the onset before 24 years of life (Kessler et al., 2007), the investigation of deviations from typical developing trajectories using machine learning methods could be a useful approach.

The question about the impact of neurodevelopmental discoveries on machine learning methods is not only difficult but also intriguing. First, in most cognitive development stages theories (e.g. Piaget), young children have low ability in complex and abstract reasoning, which improves with age and correlates with the maturational stages of the prefrontal cortex (Giedd & Rapoport, 2010). Second, many studies demonstrated that in neurodevelopment there is a joint process of functional integration and segregation of brain networks (Fair et al., 2007). Regarding the first point, it is interesting to observe that in many deep learning methods, it is frequently claimed that the level of conceptual abstraction and complexity encoding increases from the first layers to the more hidden layers. In addition, when training the Stacked Restricted Boltzmann Machines, the training starts from the first layers to the hidden layers. Moreover, segregation is associated with specialization in neural modules, while integration connects these distinct brain systems. Interestingly, the first layers of a deep belief network applied to image recognition often represent specialized feature detectors. In the more hidden layers, these specialized systems are then integrated to represent more complex features. Since prefrontal cortex is described as an integrative region and necessary for more complex reasoning, abstract thinking and decision-making, the similarity of developmental stages with deep learning training is remarkable. As a conclusion, I believe deep learning could be used to model some aspects of brain systems, and discoveries on neurodevelopment might also be useful to inspire deep architectures and training algorithms.

Bibliography

Bowers, J.S. & Davis, C.J. (2012). Bayesian just-so stories in psychology and neuroscience. *Psychological Bulletin*, 128(3), 389-414.

Brayanov, J.B., & Smith, M.A. (2010). Bayesian and 'anti-Bayesian' biases in sensory integration for action and perception in the size-weight illusion. *Journal of Neurophysiology*, 103, 1518-1531.

Chater, N. & Oaksford, M. (eds.) (2008). *The probabilistic mind: prospects for Bayesian cognitive science*. New York: Oxford University Press.

Cheng, J.; Greiner, R.; Kelly, J.; Bell, D.A. & Liu, W. (2002). Learning Bayesian networks from data: an information-theory based approach. *Artificial Intelligence*, 137, 43-90.

Clark, A. (2013). Whatever next? Predictive brains, situated agents and the future of cognitive science. *Behavioral and Brain Sciences*, 36, 181-204.

Cleophas, T.J. & Zwinderman, A.H. (eds.) (2013-14). *Machine learning in medicine: cookbook*. Vols. 1-3. Dordrecht: Springer.

Coppola, D.M.; Purves, H.R.; McCoy, A.N. & Purves, D. (1998). The distribution of oriented contours in the real world. *Proceedings of the National Academy of Sciences U.S.A.*, 95(7), 4002-4006.

Croskerry, P. (2003). The importance of cognitive errors in diagnosis and strategies to minimize them. *Academic Medicine*, 78(8), 775-780.

Croskerry P.; Petrie D.A.; Reilly J.B. & Tait, G. (2014). Deciding about fast and slow decisions. *Academic Medicine*, 89, 197-200.

Doi, K.; MacMahon, H.; Katsuragawa, S.; Nishikawa, R.M. & Jiang, Y. (1999). Computer-aided diagnosis in radiology: potential and pitfalls. *European Journal of Radiology*, 31(2), 97-109.

Dosenbach, N.U.; Nardos, B.; Cohen, A.L.; Fair, D.A.; Power, J.D.; Church, J.A.; Nelson, S.M.; Wig, G.S.; Vogel, A.C.; Lessov-Schlaggar, C.N.; Barnes, K.A.; Dubis, J.W.; Feczko, E.; Coalson, R.S.; Pruett, J.R., Jr.; Barch, D.M.; Petersen, S.E. & Schlaggar, B.L. (2010). Prediction of individual brain maturity using fMRI. *Science*, 329 (5997), 1358-1361.

Dotov, D.G. (2014). Putting reins on the brain. How the body and environment use it. *Frontiers in Human Neuroscience*, 8, 1-12.

Evans, J.S.B.T. & Stanovich, K.E. (2013). Dual-process theories of higher cognition: advancing the debate. *Perspectives on Psychological Science*, 8, 223-241.

Fair, D.A.; Dosenbach, N.U.; Church, J.A.; Cohen, A.L.; Brahmbhatt, S.; Miezin, F.M.; Barch, D.M.; Raichle, M.E.; Petersen, S.E. & Schlaggar, B.L. (2007). Development of distinct control networks through segregation and integration. *Proceedings of the National Academy of Sciences U.S.A.*, 104(33), 13507-13512.

Feher da Silva, C. & Baldo, M.V.C. (2012). A simple artificial life model explains irrational behavior in human decision-making. *PLOS ONE*, 7(5), e34371.

Friston, K.J.; Litvak, V.; Oswal, A.; Razi, A.; Stephan, K.E.; van Wijk, B.C.; Ziegler. G. & Zeidman, P. (2016). Bayesian model reduction and empirical Bayes for group (DCM) studies. *Neuroimage*, 128, 413-431.

George, D. & Hawkins, J. (2009). Towards a mathematical theory of cortical micro-circuits. *PLOS Computational Biology*, 5(10), e1000532.

Giedd, J.N. & Rapoport, J.L. (2010). Structural MRI of pediatric brain development: what have we learned and where are we going? *Neuron*, 67(5), 728-734.

Graber, M.L.; Franklin, N. & Gordon, R. (2005). Diagnostic error in internal medicine. *Archives of Internal Medicine*, 165(13), 1493-1499.

Jangholi, N. (2014). *Um sistema de rede bayesiana para diagnóstico de tinnitus.* Master's thesis, Universidade Federal do ABC, São Bernardo do Campo, Brazil.

Jones, M. & Love, B.C. (2011). Bayesian fundamentalism or enlightenment? On the explanatory status and theoretical contributions of Bayesian models of cognition. *Behavioral and Brain Sciences*, 34(4), 169-188.

Kessler, R.C.; Angermeyer, M.; Anthony, J.C.; De Graaf, R.; Demyttenaere, K.; Gasquet, I.; De Girolamo, G.; Gluzman, S.; Gureje, O.; Haro, J.M.; Kawakami, N.; Karam, A.; Levinson, D.; Medina Mora, M.E.; Oakley Browne, M.A.; Posada-Villa, J.; Stein, D.J.; Adley Tsang, C.H.; Aguilar-Gaxiola, S.; Alonso, J.; Lee, S.; Heeringa, S.; Pennell, B.E.; Berglund, P.; Gruber, M.J.; Petukhova, M.; Chatterji, S. & Ustün, T.B. (2007). Lifetime prevalence and age-of-onset distributions of mental disorders in the World Health Organization's World Mental Health Survey Initiative. *World Psychiatry*, 6(3), 168-176.

Knill, D.C. & Richards, W. (eds.) (1996). *Perception as Bayesian inference.* New York: Cambridge University Press.

McClelland, J.L. & Rumelhart, D.E. (1986). *Parallel distributed processing. explorations in the microstructure of cognition.* Vol. 2: *Psychological and biological models.* Cambridge (MA): MIT Press.

Niv, Y.; Joel, D.; Meilijson, I. & Ruppin, E. (2002). Evolution of reinforcement learning in uncertain environments: a simple explanation for complex foraging behaviors. *Adaptive Behavior*, 10(1), 5-24.

Norman, G.; Sherbino, J.; Dore, K.; Wood, T.; Young, M.; Gaissmaier, W.; Kreuger, S. & Monteiro, S. (2014). The etiology of diagnostic errors: a controlled trial of system 1 versus system 2 reasoning. *Academic Medicine*, 89(2), 277-284.

Rumelhart, D.E. & McClelland, J.L. (1986). *Parallel distributed processing: explorations in the microstructure of cognition.* Vol. 1: *Foundations.* Cambridge (MA): MIT Press.

Sato J.R.; Biazoli, C.E., Jr.; Salum, G.A.; Gadelha, A.; Crossley, N.; Satterthwaite, T.D.; Vieira, G.; Zugman, A.; Picon, F.A.; Pan, P.M.; Hoexter, M.Q.; Anés, M.; Moura, L.M.; Del'aquilla, M.A.; Amaro, E., Jr.; McGuire, P.; Lacerda, A.L.; Rohde, L.A.; Miguel, E.C.; Jackowski, A.P. & Bressan, R.A. (2015). Temporal stability of network centrality in control and default mode networks: specific associations with externalizing psychopathology in children and adolescents. *Human Brain Mapping*, 36(12), 4926-4937.

Sato, J.R.; Hoexter, M.Q.; Castellanos, X.F. & Rohde, L.A. (2012). Abnormal brain connectivity patterns in adults with ADHD: a coherence study. *PLOS ONE*, 7(9), e45671.

Schiff, G.D.; Hasan, O.; Kim, S.; Abrams, R.; Cosby, K.; Lambert, B.L.; Elstein, A.S.; Hasler, S.; Kabongo, M.L.; Krosnjar, N.; Odwazny, R.; Wisniewski, M.F. & McNutt, R.A. (2009). Diagnostic error in medicine: analysis of 583 physician-reported errors. *Archives of Internal Medicine*, 169(20), 1881-1887.

Singh, H.; Davis Giardina, T.; Meyer, A.N.D.; Forjuoh, S.N.; Reis, M.D. & Thomas, E.J. (2013). Types and origins of diagnostic errors in primary care settings. *Journal of the American Medical Association*, 173(6), 418-428.

Wasyluk, H.; Onisko, A. & Druzdzel, M. J. (2001). Support of diagnosis of liver disorders based on a causal Bayesian network model. *Medical Science Monitor*, 7 (suppl. 1), 327-332.

Wiens, J. & Wallace, B.C. (2016). Editorial: Special issue on machine learning for health and medicine. *Machine Learning*, 102, 305-307.

Winkler, A.M.; Webster, M.A.; Brooks, J.C.; Tracey, I.; Smith, S.M. & Nichols, T.E. (2016). Non-parametric combination and related permutation tests for neuroimaging. *Human Brain Mapping*, 37(4), 1486-1511.

Wozny, D.R.; Beierholm, U.R. & Shams, L. (2010). Probability matching as a computational strategy used in perception. *PLOS Computational Biology*, 6(8), 1-7.

Chapter 17

Computational models for simulating cognition and behavior: reflections on their methods, scope, and underlying epistemology

Diego Zilio
Dept. of Social and Developmental Psychology
Federal University of Espírito Santo, Vitória, ES, Brazil
diego.zilio@terra.com.br

Flávio Soares Corrêa da Silva
Institute of Mathematics and Statistics
University of São Paulo (USP), São Paulo, SP, Brazil
fcs@ime.usp.br

Luciano Silva
School of Computation and Informatics
Mackenzie Presbyterian University, São Paulo, SP, Brazil
luciano.silva@mackenzie.br

Abstract

This paper comprises three approaches to the discussion of theoretical issues related to models for the simulation of cognition and behavior. In section 1, one discusses the role of models in explaining psychological phenomena, exploring the differences and similarities of these models in comparison with experimental models developed in the biological sciences. Section 2 discusses foundational and methodological issues related to modeling cognitive phenomena, with the specific interest in embodying models in artifacts. Section 3 addresses the issue of representing cognitive concerns in computational interfaces using algebraic constructions from Computational Semiotics.

1. Explaining psychological phenomena: the role of experimental and artificial simulations

What is the role of simulation in explaining psychological phenomena? Our goal in this section is to discuss this question. However, in order to approach this subject properly, we have to discuss some other questions deeply related to it. The first one being: what kind of science is psychology? The second question, clearly related to the first, is what counts as an explanation? These are what may be called "foundational questions". Although there is no consensus about the answers to these questions, we will present a specific point of view about them from which one may start to discuss the problem of experimental and artificial models of explanation.

What kind of science is psychology? To put it simply: psychology is part of the natural sciences. More specifically, it is part of biology (Marr & Zilio, 2013; Zilio, 2016). Behavior is a biological property of organisms. Being part of biology, psychological explanations are biological explanations, which restricts our second question to the issue of what counts as explanation in biology.

There is a relatively new movement in philosophy of science that has biology as the model for discussing science. This movement is known as the "new mechanism". According to this view, the main way of doing science in biology and, by extension, proposing explanations of biological phenomena, is by describing the mechanisms related to the production of biological phenomena. When we say "biological mechanism", what do we exactly mean by "mechanism"? Bechtel & Richardson (2010, p. 17), in their now classic book *Discovering complexity* (originally published in 1993), presented the following definition: "A machine is a composite of interrelated parts, each performing its own functions, that are combined in such a way that each contributes to producing a behavior of the system". The workings of a mechanism are usually illustrated by images showing its parts (or elements), their functions in the context of the system they are part of, and how those parts relate to each other.

What counts as an explanation in biological mechanism? According to Craver (2005, p. 386):

> *The goal in describing a mechanism [...] is to describe this productive flow, without gaps from the beginning of the mechanism to its end. When we show how a mechanism works, we describe the relevant parts of the mechanism and their relevant properties, and we show how their activities are organized together so that the working of the mechanism is transparent and unmysterious.*

In other words, to explain is to describe the variables responsible for the production of the phenomenon under analysis. However, there is no such thing as explanation "without gaps". This is an idealized goal. Again with Craver (2006, p. 360): "Few if any mechanistic models provide ideally complete description of a mechanism. In fact, such descriptions would include so many potential factors that they would be unwieldy for the purposes of prediction and control and utterly unilluminating to human beings".

We can follow two paths to sustain this idea. The first is to defend that humans have no epistemological capacity to provide complete explanations of mechanisms. The second, which I think is more interesting, is a pragmatical reason. An ideally complete description of a mechanism would be so complex, would include so many variables, that it would probably be of no use for predicting and controlling the phenomena we are trying to understand. In this context, the criterion for evaluating a mechanism-based explanation is pragmatic. The description of a mechanism does not need to provide complete descriptions of the functions of all its parts in order to represent the phenomena. They need to provide knowledge to the point of making it possible to predict and control the phenomena they are explaining.

According to Glennan (2005, p. 446), explanations based on mechanical models have two main characteristics: "A mechanical model consists of (i) a description of the mechanisms' behavior (the behavioral description); and (ii) a description of the mechanism that accounts for that behavior (the mechanical description)". To describe the "behavior" of the mechanism is to describe the phenomenon we are trying to explain. By its turn, to describe the mechanism that accounts for that behavior is to describe the parts, their functions, and the interrelation between them (i.e., the organization), that results in the production of behavior. Another way of putting it is to say that the behavioral description tells us what a mechanism is doing, while the mechanical description tells us how the mechanism does it. This distinction between the description of the behavior of a mechanism and the description of the mechanism itself is paramount for understanding the differences between experimental and artificial models.

Experimental models are the ones we commonly find in basic research with humans and non-human subjects. There is a relation of isomorphism between models studied in experimental settings and the phenomena they aim to explain, which means, for instance, that if one studies a particular brain process in the laboratory, he is studying the mechanisms as they are outside the experimental setting. The difference is that one selects a few variables of interest (in a way, one "simplifies" the system), so that the experimental study can be possible. That is the reason why we talk about experimental *models*, because they only comprise some of the elements that constitute the mechanism related to the production of the phenomenon under study.

From these characteristics, we gather that experimental models aim to: (i) produce the mechanisms' behavior; and (ii) describe the mechanism that accounts for that behavior. Therefore, they comprise both aspects of Glennan's (2005) definition of mechanical explanations.

Artificial models, on the other hand, are usually carried out by computers, some sort of AI, robots, and so on. There is no structural identity between model and phenomena. Contrary to what happens with experimental models, you are not necessarily studying the mechanisms as they are outside the experimental setting. However, there can be functional similarities, which means that the elements of artificial models can produce similar (or, better saying, "analogous") functions in comparison to the elements of the phenomena they are trying to simulate (for instance, neural network models of brain function). But that is not always the case. Usually, a computer runs a simulation based on a particular algorithm related to the phenomena in question (for instance, simulations of climate changes or economical models).

Artificial models usually start in the same way as experimental models: with the definition of the phenomena under study. However, the following steps are quite different. Instead of decomposing the actual mechanism, one creates a "how-possible" model of the mechanism. The primary question is: how can I predict/produce the behavior under study? The last step is to run simulations in order to see if the model produces functionally similar behavior in comparison with the phenomena or if by using it we can predict the behavior of the phenomena in its natural settings.

From that we can gather that artificial models aim to describe, produce, and predict behavior that is functionally similar to mechanisms' behavior. It's not primarily interested in how the actual mechanism works. Therefore, artificial models do not comprise both aspects of Glennan's (2005) definition of mechanical explanations.

That is, of course, a very crude description of both experimental and artificial models. But let's analyze their characteristics in the light of what counts as an explanation in the biological sciences (including psychology). As stated earlier, the criterion for evaluating a mechanism-based explanation is pragmatic. Mechanism models do not need to provide complete descriptions of the functions of their parts in order to produce the phenomena. They need to provide knowledge to the point of making possible the prediction and control of the phenomena they are explaining. We argued that experimental models aim to predict and produce mechanism behavior by means of describing the actual mechanism that accounts for that behavior. From this second characteristic follows the means to *control* the occurrence of phenomena. We can only control biological phenomena by knowing how it actually works, by manipulating it directly. Consequently, experimental models fulfill the pragmatic criteria of explanation.

On the other hand, artificial models aim to describe, produce and predict behavior that is functionally similar (or analogous) to mechanism behavior. It's not the mechanism behavior per se. Artificial models are not primarily interested in how the actual mechanism works. In light of the pragmatic criteria (that is, prediction and control), artificial models can only contribute to the prediction of phenomena by means of simulating the occurrence of functionally similar behavior. Take, for instance, artificial models of climate change. Those models do not entail direct manipulation of the environment. However, by using artificial models we can simulate the occurrence of some results by adding variables that we think might influence climate change. And then we can predict that, if those conditions occur in the real environment, probably some consequences will follow, because that is what the simulation tells us.

Artificial models do not explain how the actual mechanism works. They are very useful, however, in giving possible predictions. Therefore, as a consequence, we should not say that artificial models promote explanations of biological or psychological phenomena. After all, they do not fulfill the criteria of prediction and control.

Of course, that does not mean they are useless. Artificial models are heuristic tools. Artificial simulations play an important role in the process of developing hypotheses, models or even new theories. Simulations may help scientists to explore situations that cannot momentarily be investigated by experimental means. They can also be used when the performance of an experiment is not possible for pragmatic, theoretical or ethical reasons.

In sum, experimental models are essential for the construction of psychological knowledge and must precede artificial models when possible, since we can only fulfill the pragmatic explanatory criteria (prediction and control of phenomena) through direct manipulation. Artificial simulations, on the other hand, have at least two main functions: to contribute to the validation of the knowledge produced by experimental work and to create useful technologies aiming at the resolution of human problems. We could say they are complementary strategies, although in biology (and experimental branches of psychology) experimental models are more commonly used, probably because of the pragmatic grounds on which its conception of explanation is founded.

2. On the different sorts of artificial intelligence: deep, shallow and mimicking

The field of Artificial Intelligence (AI) as we know it today started in 1956 in the United States, as an initiative of John McCarthy and colleagues (Moor, 2006). It was planned as a multidisciplinary field and has always aimed at the study of Intelligence by means of its reconstruction in human designed platforms. Around the same period and location, in the U.S., the field of Human-Computer

Interaction (HCI) started to be structured around the notion of *human augmentation* proposed by Douglas Engelbart, among other scholars (Engelbart, 1988). For several years, AI and HCI were developed by separate, nearly disjoint research communities, carrying conflicting views about the most appropriate ways of bringing together humans and digital/computational machines.

All along the history of AI, some initiatives have taken the road of foundational scientific endeavor, while other have focused on the appropriation of techniques inspired on biological phenomena to design and build useful artifacts. Within the AI research community, some scholars have named the former initiatives *Deep AI,* and the latter *Shallow AI.*

Deep AI refers to attempts and initiatives to isolate Intelligence as an observable phenomenon, and to build a deeper and broader understanding of this phenomenon through model building and simulation in man-made substrates.

Shallow AI, in contrast, emphasizes the potentialities of techniques that emerge as a result of observing biological systems whose behavior can be deemed *intelligent,* and which can be used as tools for building artifacts that, based on some appropriate metric, can be considered an improvement over previously existing artifacts of similar functionality.

A third possible road, in which digital artifacts could be designed to convincingly *mimic* intelligent behavior, was initially viewed as a road to be avoided, as the results of initiatives built this way could be taken as forgery or potentially unethical behavior. Recently, however, interesting methodological developments have taken place, aligning the fields of AI and HCI and clarifying that Deep, Shallow and Mimicked AI were not, in fact, as different as initially considered (Winograd, 2006).

By taking a more encompassing (and hence significantly more challenging) design stance, one can consider the design of social systems comprised by digital/computational artifacts as well as human participants. Such systems – frequently termed *Social Machines* – should be structured in such a way that digital components are programmed and human components are incentivized effectively to cooperate (Robertson & Giunchiglia, 2013).

In order to build high quality social machines, one needs to build appropriate interaction networks and protocols, program the behavior of digital components based on well grounded (*deep*) models of intelligence, make sure that these models are computationally efficient (hence implemented according to the precepts of good *shallow* models of intelligence), and, finally, make sure that digital components can be perceived by human components as intelligent, so that social interactions can occur with the required fluidity. As a consequence, *mimicked* intelligence has been accepted as a third facet of intelligence that is required to build digital devices which can deserve to be accepted as intelligent in social interactions.

3. Algebraic semiotics for specification of cognitive aspects in human-computer interaction

Computational Semiotics (Gudwin & Queiroz, 2005) is an interdisciplinary field that draws on research in Logic, Theory of Computation, Formal and Natural Language Processing, Cognitive Sciences, and Semiotics proper. A common theme in Computational Semiotics is the adoption of a sign-theoretic perspective on issues of Artificial Intelligence and Knowledge Representation. One important application of Computational Semiotics lies in the field of Human-Computer Interaction (HCI).

Algebraic Semiotics is a part of Computational Semiotics, which combines aspects of Algebraic Specification and Social Semiotics, and it has been applied to the specification, implementation and verification of assets and processes in user interface design (Goguen, 1999).

According to Goguen (1999), the results in HCI are either specialized and precise, or general but of uncertain reliability. One can obtain information on these facts in protocol and usability analyses, questionnaire results and case studies. On the other hand, with the evolution of techniques, devices and processes in HCI, one needs a concise theory to support interface design and adapt it to new media, metaphors, devices and users with special needs (e.g. users with disabilities).

For these new scenarios, Algebraic Semiotics provides formal (algebraic) definitions based on Algebra, by using a special signal system and representation. Moreover, it presents a calculus of representation and rules for combining representations. And, most importantly, Algebraic Semiotics provides precise ways of comparing the quality of representations.

Algebraic Semiotics uses a sign system, which is composed of:

- signatures for sorts, subsorts and operations (constructors and selectors)
- equations for modeling constraints
- data sorts and functions
- levels for sorts
- priority ordering on operations

Goguen (1999) points out that sorts classify signs, operations construct signs, data sorts provide values for attributes of signs, and levels and priority indicate saliency. Despite this definition not being very formal, it is powerful enough to model various elements in HCI .

User interface design, essentially, deals with the production of good representations. For example, Graphical User Interface (GUI) represents functionality with buttons, icons, menus, windows, among others. So, the first point to be considered is how to formalize representation in HCI.

Goguen (1999) proposes the formalization of a representation as a map $R: S_1 \rightarrow S_2$ of sign systems, called semiotic morphisms. These morphisms should preserve sorts and subsorts, operations, axioms, data, functions, levels of sorts and priority of constructors.

Let us consider S_1 as a signal system formed by English sentences and S_2 the parse trees of English Sentences. One example of an action of a semiotic morphism R is depicted in Fig. 17.1:

```
                                          S
                                         / \
                                       NP   VP
                                      /    / \
                                     N    V   PP
time flies like an arrow     →      /     |   / \
                                 time flies  P   NP
                                            /   / \
                                          like an  arrow
```

Figure 17.1: A semiotic morphism mapping an English Sentence onto its parse tree (adapted from Goguen, 1999, p. 253).

In the parse tree, S represents a label for Sentence, NP is a noun phrase, VP is a verbal phrase, N is a noun, V is a verb, PP is a prepositional phrase and, finally, P is a preposition. For more details on algebraic property preservation, see Goguen (1999).

Considering a text-based user interface, it is possible to recognize certain cognitive aspects, such as emotions and moods (Calvo et al., 2014). Typical examples of emotional states are nervous, tired, happy and bored. A typical example of mood is depression, which can last several days. Emotions are easier to detect because they are shorter than moods and there are several sensors to assess their indirect signs.

Considering the representation problem of the emotional cognitive state 'nervous', the first step is to build a finite set multimodal X of sorts (Silva, 2015):

$$X = \{x^{face}, x^{voice}, \ldots, x^{pulse}\},$$

where each sort x is a random variable of a modality which is measuring the emotional state ω. The measure of this state is defined by:

$$P(\omega \mid X) .$$

At the same time as this probability is a measure, it is also an algebraic representation of the emotional state 'nervous'. This representation states that one can estimate the emotional state given the probabilities that the sensors are in a particular state. For example, assuming that x^{face} is the probability that the user's face is flushed, x^{voice} the probability that the voice is shaky, and x^{pulse} the probability that the pulse is accelerated, two instances of the measure of the state are:

$$P(nervous \mid x^{face} = 0.9, x^{voice} = 1.0, x^{pulse} = 0.9) \text{ is high (about 1)},$$

$$P(nervous \mid x^{face} = 0.1, x^{voice} = 0.0, x^{pulse} = 0.15) \text{ is low (about 0)}.$$

In order for the user to be in the emotional state of 'nervous', it is necessary that the values of some of the modalities be above a certain value. Suppose now that a single modality is being measured. Using the well-known definition of joint probability as a function of conditional probability, one has:

$$P(\omega, x^{face}) = P(\omega)P(x^{face} \mid \omega) ,$$

$$\text{so that } P(\omega \mid x^{face}) \propto P(\omega, x^{face}) .$$

Taking two *independent* sensors, x^{face} and x^{pulse}, it is possible to make the following approximation:

$$P(\omega \mid x^{face}, x^{pulse}) \propto P(\omega, x^{face}, x^{pulse}) = P(\omega)P(x^{face} \mid \omega)P(x^{pulse} \mid \omega),$$

$$\text{so that } P(\omega \mid x^{face}, x^{pulse}) \propto P(\omega \mid x^{face})P(\omega \mid x^{pulse}).$$

The latter equation states, essentially, that the probability of a user being in the emotional state ω is approximately the product of the probabilities of the states of the two sensors. It can also be interpreted as an algebraic product of the single modality representations $P(\omega | x^{face})$ and $P(\omega | x^{pulse})$.

Generalizing this idea for *n* modality sensors, one has:

$$P(\omega | x^1 ... x^n) \propto P(\omega | x^1) ... P(\omega | x^n).$$

All the presented equations may be used to model restrictions in emotional user interfaces. It is also possible to define levels for each of the sorts: for example, one can make measurements of the faces and then of the pulse and voice. This same process may be used as an ordering of measurements (Silva, 2014).[1]

[1] Section 1 was written by Diego Zilio, section 2 by Flavio Soares Corrêa da Silva, and section 3 by Luciano Silva.

Bibliography

Bechtel, W. & Richardson, R. C. (2010). *Discovering complexity: decomposition and localization as strategies in scientific research.* Cambridge (MA): MIT Press. Original: 1993.

Calvo, R.A.; D'Mello, S.; Gratch, J. & Kappas, A. (2014). *The Oxford handbook of affective computing.* Oxford: Oxford University Press.

Craver, C.F. (2005). Beyond reduction: mechanisms, multifield integration and the unity of neuroscience. *Studies in History and Philosophy of Biological and Biomedical Sciences,* 36, 373-395.

——— (2006). When mechanistic models explain. *Synthese,* 153, 355-376.

Engelbart, D.C. (1988). The augmented knowledge workshop. In: Goldberg, A. (ed). *A history of personal workstations.* Boston: Addison-Wesley, pp. 185-236. Available at: https://www.dougengelbart.org/pubs/augment-101931.html

Glennan, S. (2005). Modeling mechanisms. *Studies in History and Philosophy of Biological and Biomedical Sciences,* 36, 443-464.

Goguen, J. (1999). An introduction to algebraic semiotics, with application to user interface design. In: Nehaniv, C.L. (ed.). *Computation for metaphors, analogy and agents.* Lecture Notes in Computer Science, 1562. Berlin: Springer, pp. 242-291. Available at: https://cseweb.ucsd.edu/~goguen/pps/as.pdf .

Gudwin, R. & Queiroz, J. (2005). Towards an introduction to computational semiotics. *Proceedings of the 2005 IEEE International Conference on Integration of Knowledge Intensive Multi-Agent Systems – KIMAS'05.* Waltham (MA): IEEE Press, pp. 393-398.

Marr, M.J. & Zilio, D. (2013). No island entire of itself: reductionism and behavior analysis. *European Journal of Behavior Analysis,* 14, 241-257.

Moor, J. (2006). The Dartmouth College Artificial Intelligence Conference: the next fifty years. *AI Magazine,* 27(4), 87-91.

Robertson, D.S. & Giunchiglia, F. (2013). Programming the social computer. *Philosophical Transactions of the Royal Society A,* 371(1987), 1-14.

Silva, L. (2014). Computação sêntica: fundamentos e aplicações em RV/RA. *Tendências e Técnicas em Realidade Virtual e Aumentada,* 4(1), 82-102.

——— (2015). Self quantificado e a integração de dados cotidianos de usuários em aplicações para RA. *Tendências e Técnicas em Realidade Virtual e Aumentada,* 5(1), 50-69.

Winograd, T. (2006). Shifting viewpoints: artificial intelligence and human-computer interaction. *Artificial Intelligence,* 170(18), 1256-1258.

Zilio, D. (2016). On the autonomy of psychology from neuroscience: a case study of Skinner's radical behaviorism and behavior analysis. *Review of General Psychology,* 20(2), 155-170.

Chapter 18

Advances and perspectives in cognitive architectures

Angelo Loula
Department of Exact Sciences
State University of Feira de Santana (UEFS), Feira de Santana, BA, Brazil
angelocl@uefs.br

Marcio Lobo Netto
Polytechnic School
University of São Paulo (USP), São Paulo, SP, Brazil
marcio.netto@usp.br

Mauro Muñoz
Polytechnic School
University of São Paulo (USP), São Paulo, SP, Brazil
mesmunoz@usp.br

Ricardo Gudwin
School of Electrical Engineering and Computation
Campinas State University (Unicamp), Campinas, SP, Brazil
gudwin@dca.fee.unicamp.br

Abstract

The paper presents four fundamental questions concerning cognitive architecture, with answers provided by the four authors. What exactly does the word "cognitive" mean? What is the difference between a cognitive system and a non-cognitive intelligent system? Is language a necessary feature of cognitive systems? What are the frontiers of research regarding cognitive architectures?

1. Introduction

Cognition is an open concept with no clear boundaries. The word "cognitive" has become a hype, and big companies like Microsoft, IBM, Google and others are now providing some sort of "cognitive" software, but it is not really clear what this means. Cognition is studied in various fields with different perspectives, including psychology, philosophy, neuroscience, linguistics, biology, social sciences, artificial intelligence and computer science, so, from a scientific perspective, to study cognition, one should start defining one's theoretical framework and principles.

Those pursuing research on cognitive architectures and systems have to deal with various perspectives and open questions. A starting point is making a distinction between natural and artificial systems. Study of natural cognitive systems aims at getting a better and deeper understanding of their fundamental aspects. As one of the consequences of the acquired knowledge on this matter, we foresee the possibility of starting to think about synthetic cognitive systems, or those designed by humans with a practical or theoretical purpose, while pursuing the fundamental aspects that define a system as cognitive.

The authors discuss relevant and fundamental questions concerning cognitive systems research, addressing the attribute "cognitive" in "cognitive systems", differentiating cognitive system and non-cognitive intelligent systems, discussing language as an important feature in a cognitive system, and debating the research frontiers on cognitive systems and cognitive architectures.

2. Questions

2.1 What does the attribute "cognitive" in "cognitive systems" refer to?

Loula: To describe the relation of the attribute "cognitive" with the concept of "cognitive systems", it would be necessary to first discuss what "cognition" is. Cognition is related with mind processes, but it is a concept with no clear boundaries and with no consensual definition. The following excerpts illustrate this:

Question 1: What is cognition?

Answer: Information processing as symbolic computation – rule-based manipulation of symbols.

Question 2: How does it work?

Answer: Through any device which can support and manipulate discrete functional elements – the symbols. The system interacts only with the form of the symbols (their physical attributes), not their meaning.

Question 3: How do I know when a cognitive system is functioning adequately?

Answer: When the symbols appropriately represent some aspect of the real world, and the information processing leads to a successful solution of the problem given to the system.

Question 1: What is cognition?

Answer: Enaction: A history of structural couplings that brings forth a world.

Question 2: How does it work?

Answer: Through a network consisting of multiple levels of interconnected, sensorimotor subnetworks.

Question 3: How do I know when a cognitive system is functioning adequately?

Answer: When it becomes part of an ongoing existing world (as the young of every species do) or shapes a new one (as happens in evolutionary history).

Both excerpts come from the book *The embodied mind: cognitive science and human experience,* by Francisco J. Varela, Eleanor Rosch & Evan Thompson (1991, pp. 42-43, 206-207), discussing different views on cognition. The first one comes from Cognitivism (Cognitive Psychology), which views the mind as an information processing unit, and the second definition is from Embodied (Enactive) Cognition, which recognizes the mind as tightly coupled with the body and with the interaction in an environment. Thus, to relate “cognitive” with “cognitive systems”, one should start defining their theoretical framework and principles.

Moreover, since cognition is studied in various fields with different perspectives, including psychology, philosophy, neuroscience, linguistics, biology, social sciences, artificial intelligence and computer science, the field of study should also come forward. The object of study should also be defined, as the variety of cognitive processes and related phenomena is also wide and may encompass topics such as knowledge, reasoning, belief, language, brain, mind, learning, adaptation, autonomy, intelligence, memory, perception, and also creativity, emotion, communication, altruism, causation, information, epistemology, representation, and more.

For the attribute “cognitive” to relate to the concept of “cognitive system”, there is a broad set of theoretical references, fields of studies and possible processes and elements of interest. Additionally, a cognitive system may be a natural system or an artificial system, and to relate the attribute ‘cognitive’ with an artificial (cognitive) system, we also have to discuss the ontological (or the epistemological) status of this relation.

An artificial cognitive system is, first of all, a system, a set of somehow interrelated elements, but it is also an artifact, designed by a human being to somehow perform a cognitive task. Borrowing a discussion on Artificial Life from a paper by H. Pattee (1989), we can inquire whether this artificial system simulates a cognitive process or whether it realizes a cognitive process. Does an artificial cognitive system function as a (literal) implementation of cognitive processes or does it (only) model and simulate such processes? Moreover, we can also ask whether artificial cognitive systems are limited to implementing or modeling natural cognitive systems and processes. Could one go beyond, from cognition-as-we-know-it to cognition-as-it-could-be? We will leave these questions open...

Netto/Muñoz: “Cognitive” is related to a general set of features that should be pursued by a system taking this name. “Cognitive” means the ability to express or to possess cognition, which means reasoning in its broadest sense, not only logical reasoning, but any kind of sensorial interpretation leading to appropriate actions.

“Cognitive systems” refers, as the name suggests, to systems with cognitive capacities. We differentiate between natural and artificial systems. The study of natural cognitive systems aims at getting a better and deeper understanding of their fundamental aspects, including for instance how it is structured, how it develops in the natural neuronal substrate, and so on. This knowledge can be viewed from two main perspectives. On the one hand, supported by neuroscientific analysis, and on the other, by pure philosophical reasoning. As one of the consequences of the acquired knowledge on this matter, we foresee the possibility of creating synthetic cognitive systems, or those designed by humans with a practical or theoretical purpose, while pursuing the fundamental aspects that define a system as cognitive.

It seems that in order to be properly addressed as "cognitive", an artificial system should meet certain expectations, which from a general point of view would be to reflect the corresponding natural systems, keeping their main features. Once we are able to identify and to (at least partially) describe the features of a cognitive system, we should also be able to build artificial ones embedded with such cognitive features. Many questions arise from this expectation, and here we try to identify some of them.

First, natural cognitive systems are autonomous and, if we intend that artificial systems share this attribute, we should not interfere on the way cognition arises in this substrate. The question is how to build a substrate with such independence, where cognition (even the most basic one) can arise. Second, natural cognitive systems result in fact from a strong interrelation among its many component subsystems, in a highly distributed form, making it difficult to understand how cognition emerges. An artificial cognitive system should also have these properties, with autonomy.

So, concluding, more than stating a list of cognitive attributes, expected to be found on each system recognized as such, it is important to understand the main fundamental aspects of a natural cognitive system, and to implement them in an artificial one, giving the system all the autonomy it may have. Of course, the tricky point is how to program such a self-evolving system so that it will perform the expected things that we humans would recognize in a cognitive system. We thus see that there are still many open questions.

Gudwin: In principle, I agree both with Loula's and Netto/Muñoz' views. We are currently seeing an explosion in the use of term "cognitive" in computer and systems science. Big companies like Microsoft, IBM, Google and others are now providing some sort of "cognitive" software, and it is not really clear what this means. The word has become a hype, without a clear understanding of its scope and limitations. As Loula already pointed out, the word "cognitive" has been used extensively under the domain of *Cognitive Sciences*, and even there with different meanings, depending on whether the researcher is a cognitivist, a connectionist, a dynamicist, or an embodied and situated enactivist.

The term "cognitive" appears throughout the scientific literature. We can see expressions like "cognitive systems", "cognitive radio", "cognitive computing", "cognitive management", "cognitive robotics", etc., and one particularly interesting question is what the authors have in mind when they employ the term "cognitive" to describe their realizations. There are many technologies which are pointed out as being "cognitive", such as Symbolic Rule Based Reasoning, Neural

Networks (Reinforcement Learning, and more recently Deep Learning), Speech Recognition, Natural Language Understanding, and many others.

Browsing the literature brings us a mix of many different meanings for this term. In many cases, the authors employ a completely non-technical, or naïve use of this term: a use without any technical explanation for its meaning, as if it were self-evident or self-explanatory for the reader (e.g. Dorneich et al., 2005; Xu et al., 2014; Foteinos et al., 2013; Vlacheas et al., 2013). In other papers, the authors implicitly define "cognitive" as related to some sort of sensing/actuating capability: used to define systems with sensing abilities and actuating on the environment (e.g. Gaussier, 2001; Hershberg & Efroni, 2001). Finally, we find other authors who use the term "cognitive" with reference to some sort of cognitive modeling: relying on some model of a high-level human cognitive function which is emulated/simulated in the artificial system (e.g. Dodd & Gutierrez, 2005; Riegler, 2002). Different models of high-level human cognitive functions, like perception, attention, decision-making, planning, action-selection, behavior generation, learning, memory, emotions, language use, consciousness, imagination, social cognition, meta-cognition, etc., can then be used to derive some sort of "cognitive" system. In our opinion, the naïve use should be simply disconsidered, and the sensor/actuator use should also be avoided, as it makes clearly a misuse of the term: any control system might have sensing and actuating abilities, without necessarily being "cognitive" because of that. We conclude that the best way of understanding this term is in the high-level human cognitive functions modeling. So, we propose the following meaning for the term "cognitive": we define a cognitive system or module, or sub-system, or whatever else, as a system, module, sub-system, etc. which aims at explicitly modeling human cognitive functions in its inner mechanisms and/or algorithms.

But this is just half of the solution, because now we have the burden of understanding what is a human cognitive function. Let us be a little more systematic on this point. This is a very dangerous subject, because we are supposed to apply a terminology which was designed originally to refer to the human mind, but we intend to make a technical use of the term, so we must be prudent here. Our idea is to start with a terminology referring to a specific phenomenon which is the human mind, find a model for explaining the meaning of these terms, perform an abstraction (or a metaphor) to simulate/emulate a mental process as a computational process, and implement this computational process in an artificial system. In this manner, a human cognitive function will translate into a computational "cognitive" process, instantiated in an artificial "cognitive" system. But then, there is a problem related to the scope of what is meant to be "cognitive". In Cognitive Science (mainly in psychology), an older view used to split the mental phenomena in three different categories: perception, cognition and action. Following this older vision, both Perception and Action should be excluded from what might be accounted as "cognitive". This was the rule during the cognitivist period in Cognitive Science. As soon as the enactivists started to stress the importance of the environment in cognitive phenomena, the word "cognitive" gained a wider scope, and all mental phenomena, including perception and action, started to be incorporated in its meaning. With this, phenomena like vision, categorization, speech recognition and behavior generation started to be assumed "cognitive" as well.

2.2 What would be the difference between a cognitive system and a non-cognitive intelligent system?

Loula: To answer the question of the difference between cognitive and non-cognitive intelligent systems, we will focus on the domain of artificial systems, thus comparing artificial cognitive systems and artificial (non-cognitive) intelligent systems. As cognition was defined as an open concept with no clear borders, the same goes for intelligence, which is even more open. But if we are discussing artificial intelligence systems, this brings forward Artificial Intelligence.

The term Artificial Intelligence was coined in 1956, in the name of the Dartmouth Summer Research Project on Artificial Intelligence Conference. Before that, research in the area was not known as Artificial Intelligence, neither were intelligent systems called by this name. Intelligent systems, by that time, included programs for playing games like checkers, logic theorem provers and mathematical problem solvers. But the field of Artificial Intelligence evolved since then, and there is now a wide set of Intelligent Systems, including data mining and machine learning systems, control systems, robotic systems, and more.

Artificial Intelligence is now a broad area that covers a lot of different research agendas. Both Cognitive Systems and (non-cognitive) Intelligent Systems can be considered parts of Artificial Intelligence research efforts, but with different approaches and ambitions. Research on Artificial (non-cognitive) Intelligent Systems are focused on developing flexible and adaptive systems inspired on intelligent (natural) processes and elements, and also on other biological processes that can reveal promising algorithmic solutions to diverse problems. The aim of building Intelligent Systems is to come up with technological artifacts that solve practical problems, inspired by certain approaches. Such Intelligent Systems include (but are not limited to) neural networks systems, fuzzy systems, evolutionary computation systems, natural computing systems, bioinspired systems and robotics, and more.

Research on Artificial Cognitive Systems, on the other hand, has focused on modeling cognitive processes in an effort to reproduce more closely the relations and elements of cognition. Cognition is not just an inspiration for building systems, the aim is to simulate (or realize) cognition, most often a certain aspect of it, with scientific or technological motivations.

Netto/Muñoz: The well-known cognitive systems are the natural ones, embedded in evolved live organisms. But even in this scenario it is hard to establish the threshold above which systems should be accepted or identified as cognitive. One way of thinking would be to characterize a system by a set of (acknowledged) features that seems to be important or fundamental in cognitive systems, even if with different degrees of presence, and then to assess how present they are in each evaluated being.

In this paper, the focus is on artificial systems. For these, the main difference between cognitive and (non-cognitive) intelligent systems is that cognitive systems are closer to the natural systems than their purely intelligent counterparts. Intelligent artificial systems are normally those built upon principles of logic, in accordance with the fundamentals of classic Artificial Intelligence. We may consider intelligent systems as all those that possess intelligent abilities, independently of whether these are explicitly designed by humans (that know what they mean and how to express them algorithmically) or by any another approach leading to an autonomic development. The most important thing would be to recognize such system as intelligent after observing their behavior (depending on how they behave or solve assigned problems, humans assess them as more or less intelligent).

On the other hand, cognitive systems may also present intelligent abilities, but the way they develop is different: no human interference should be allowed in the design of their behavior, meaning they should not have any designer embedded feature. Furthermore, it is expected that the intelligence should be broader, in the sense that it does not target one specific aspect, but rather should be a general intelligence, able to solve whatever is necessary along the existence of the agent embedded with such cognitive capability. This also leads to the concept of "general learning ability".

To wrap it up, the most important difference between cognitive and non-cognitive intelligent systems is the way their intelligence is developed (natural or artificial evolution) and how it effectively works supporting the agent in which it is embedded (assisting softly on every issue, and focusing hard on specific ones).

Gudwin: I would like to insist here on the issue of cognitive modeling. It is clear that a cognitive system is a kind of intelligent system. But in my point of view, what distinguishes a cognitive system from any other intelligent system is its origin in some sort of cognitive modeling. And cognitive modeling, as I have already pointed out, refers to modeling some mental ability. In other words, if an intelligent system is developed without the support of any cognitive model, then it should not be called cognitive. This is the case, when we use standard techniques from computational intelligence, like fuzzy systems, neural networks (in a more general sense), or genetic algorithms. I would not classify these as cognitive systems, unless a clear inspiration from a cognitive model is addressed. Two topics which are receiving an increased interest in the literature are the topics of "Cognitive Architectures" and "Artificial General Intelligence". Both topics are somewhat intermixed. In both cases, the idea is to be able to construct computational architectures inspired on cognitive models (in the case of artificial general intelligence, with an extra goal of achieving human level intelligence). Both approaches use different kinds of methodologies. There are symbolic cognitive architectures like SOAR or ACT-R, neural cognitive architectures like Clarion, Leabra, and also other approaches like LIDA, OpenCog, Sigma, etc., using conceptual graphs, semantic networks, belief networks and other kinds of representational structures. There are cognitive architectures like SAL (Synthesis of ACT-R and Leabra), or maybe also Clarion, which try to hybridize symbolic and neural components in order to create a cognitive architecture which may model more closely the many facets of mental activity. There are specific conferences, nowadays, like the NeSy symposiums, running annually since 2005[1], which are trying to investigate how symbolic and neural approaches can be mixed together. So, to me the demarcation line between a cognitive and a non-cognitive system is quite clear: the existence of a cognitive model supporting the technological development. If there is a cognitive model to support it, it can be called cognitive. If there is not, then it should be non-cognitive.

2.3 How is "language" important while considering a cognitive system; is it a necessary feature, or a bonus?

Loula: Language is a remarkable feature of the human species. It is organized, generative and recursive. It is able to express and describe endless possibilities. Language qualitatively differentiates humans being from other animals. From a cognitive perspective, there are components underlying language that are uniquely human, while there are also other components, upon which language is based, that are shared by other animals, as Hauser, Chomsky & Fitch (2002) assert in their article "The faculty of language: what is it, who has it, and how did it evolve?". Language acquisition and use rely on various other cognitive competences, specific or not, and empirical studies and theoretical descriptions have been trying to understand and explain this. So, cognitive competences involved in the acquisition and use of language remains an open issue.

Language is certainly an important element to consider in research of (artificial) cognitive systems, as it remains a great challenge in the field. Current (computational) artificial systems do not handle natural language in the same way as natural systems do. Artificial systems are not able to understand what we say in the same way another person would do. In part, this difficulty exists

---

[1] See: http://www.neural-symbolic.org/

because this natural phenomenon is not satisfactorily understood, undermining an appropriate specification in a machine. However, multiple dimensions of language are not adequately considered by engineers of artificial systems, who tend to focus on the syntactical dimension. Natural language involves syntax, but also semantics and pragmatics, and it includes social, cognitive, biological and semiotic aspects.

Language is an important element but not a necessary one to be considered in further advances. There are a lot of cognitive competences, not dependent on language, that are still challenging researchers in cognitive systems. On the other hand, language can be seen as a complex communication system, and as such, a possible and scientifically adequate approach is to focus initially on simpler communication systems, before moving on to language.

Netto/Muñoz: Language seems to be a very important characteristic of a cognitive system, independently of the level it may be expressed. From simple signals communicating facts among animals, through those embedded with some level of structure and therefore with higher power to organize thoughts, to the most developed levels in human beings, in all cases language principles play an important role. But it should not be necessarily considered as a fundamental requirement for a cognitive system. In other words, we should consider that some natural systems (animals) do have some level of cognitive capacity, even though they don't express language abilities. In these cases, the cognitive capacities are very poor or constrained, due to the lack of mechanisms that socially structure and develop cognitive concepts.

This means that the "truly" cognitive systems are those enhanced with some sort of language skills. And from this we infer that language is deeply interrelated with cognition, being one of the most important or distinctive features associated to cognition. We can also say that language builds the soft substrate upon which cognition arises (or at least the most sophisticated forms of cognition).

Language is important in two ways. First because it provides a support for reasoning, which is a fundamental aspect of cognitive systems; thus, language constitutes a basis for cognitive operations. Secondly, because language helps with communication, allowing information and knowledge exchange between individuals, providing a common basis to establish references for understanding the world they live in.

Gudwin: I think that language is one of the highest cognitive abilities found in human beings. So, clearly, the abilities of understanding and producing language should be considered as important functionalities of cognitive systems. It is clearly what might become an advantage, from an external point of view (for those observing a cognitive system in action). Language may be seen as the mark of cognitive systems, together with perception (vision and speech recognition). From a functional point of view, perceptual and linguistic abilities are being explored by some software companies as the distinctive marks of cognitive systems. But even if one recognizes them as important, I don't think they are a necessary feature of a cognitive system, they are more like a bonus, together with other cognitive abilities, like learning, emotions, reasoning, memory and others. The whole issue of cognitive systems is to make artificial systems more close to our humanity, step by step, giving them abilities which on the past were exclusive to animals and human beings.

Addressing now specifically linguistic understanding, it is important to distinguish two different categories of research. The field of research of simulation of language evolution is bringing us many advances of which some people might not be fully aware. The problem of recognizing and understanding isolated words is being addressed for already some time, with many kinds of language games and many successful computational implementations. This problem can be considered already solved, from a technological perspective. But a much more complicated problem, the recognition of full sentences, conversations and dialogues, is only now being addressed in a more comprehensive way. Research is just starting regarding this. Even though some results have been reported, these results are still very preliminary. This is a topic which will be addressing attention from now on in cognitive systems, and due to that, the topic of language gains a high prominence in cognitive systems studies.

2.4 How do you see the frontiers of research regarding cognitive systems and cognitive architectures; what are the missing gaps; what are the most critical issues?

Loula: There are still a lot of open issues and possible directions for expanding research frontiers in cognitive systems. Concerning especially artificial cognitive systems, there are particular challenges in building systems that models' cognitive processes, and a challenge of great interest are representations.

Representations are a topic of interest in Artificial Intelligence (AI) since its foundation and remains an important issue in current research. The initial conception that intelligent systems are capable of reasoning based on representations, following a formal logic approach to cognition, brought about an ontological question, concerning what such representations would be, an ontological issue, and an epistemological question, on how they could be produced and interpreted. The first problem was answered by determining the appropriate data structures, in a merely technical perspective, and determining how to collect and insert data that would represent the knowledge on which inferences would be applied and new knowledge would be obtained. This led to several criticisms, such as the Symbol Grounding Problem, that essentially challenged how something could actually represent something else for an intelligent system, and not only for the designer that provided the data to the system.

On the other hand, the so called Nouvelle AI proposed a new approach for intelligent systems, committed to situatedness and embodiment of cognition. In these new systems, embodied artificial agents are situated in an environment, establishing sense and act loops, and interacting with other agents. Agents would build their cognitive competences as a consequence of their history of sensory-motor cycles and interactions, based on learning, adaptation and evolution. Nevertheless, there was a refusal to deal with representations in this new approach, maybe considered as a minor or unnecessary trait.

Meanwhile, more recently, there has been a great variety of research on the emergence of communication and language among artificial agents, robotic and simulated ones. As a methodological principle, the cognitive or social process of interest is not previously present in a community of agents, but by means of interactive and adaptive processes it can emerge among the agents. But, even though communication and language are strongly related to representational processes, there has been little or no discussion on this issue in such research reports.

Based on the fact that communication can be seen as the production (by a speaker) and the interpretation (by an interpreter) of representations, it is fundamental to understand the characteristics and conditions for the emergence of diverse modalities of representational processes, associated with communication and their relation to other cognitive traits. As such, we have been developing experiments to study underlying representational process on the emergence of communication in a community of artificial agents as a particularly relevant framework. As an essential part of our approach to study representations in communicative cognitive agents, we have used theoretical principles from Charles Sanders Peirce's semiotics to define the model of representation processes, their variety and constraints on dynamics. We believe, and our results have shown, that this is a fruitful approach for synthesizing and analyzing representation processes in artificial cognitive systems, thus expanding the frontiers of cognitive systems.

Netto/Muñoz: The areas of cognitive systems and cognitive architectures should converge more and more, since they express perhaps just different approaches to handle the same subject. We consider cognitive systems as a more general concept (including natural and artificial ones), so it should be an object of a study aiming to provide a deeper understanding of its core features (what it means to be cognitive). On the other hand, cognitive architectures correspond to an engineering approach to support the conception of synthetic cognitive systems. They therefore refer to components and their systemic interrelation, which as a whole constitute a cognitive system.

One of the missing gaps is that there is still a strong human interference on the cognitive architecture design. Normally architectures are designed with some purpose, and even if on runtime we do not see human interference anymore, it is somehow embedded in the designed system. So, it is very hard to have a full blown cognitive architecture, if one expects to have a self-evolved system.

Another important gap refers to the distance between all these artificial systems and real cognitive ones (a distance that will continue for a long time). On the other hand, one sees a continuous improvement, and there is the possibility of starting to use artificial cognitive agents as truly assistants in some fields. Perhaps this should be the way to promote the evolution of studies, projects and products: finding specific niches where this assistive technology can really bring good results.

Gudwin: The field of cognitive systems is actually at the upfront of research in artificial intelligence. There are many open fronts of research, and many important topics which deserve a deeper exploration. Among them, a prominent one is *learning*. Great progress is being reported in the field of *deep learning* in convolutional neural networks, which are being used within cognitive architectures to represent new models of *perception*, applied in computer vision and speech recognition. The field of memory, and specifically *episodic memory*, is also receiving interesting contributions. Nowadays, episodic memory is one of the important missing cognitive abilities in artificial systems, and it is receiving a lot of attention. Episodic memory is also related to the field of *machine consciousness*, which is also at the center of the debate in specialized meetings on cognitive architectures. All of these are an important background for the studies on *grammatical language*, the ability to understand full sentences, as pointed out earlier. The topic of *social cognition* is also relevant in this sense, as many systems are dealing with the real-time dialogue between humans and systems, together with vision and the recognition of the intention to dialogue from the part of human subjects. Other topics as *meta-cognition* and *computer imagination* are also being addressed, together with the more traditional topics of *planning* and *reasoning*, now being considered together with perception and action in order to compose full cognitive architectures. Finally, the topic of *emotions* is also receiving great attention, either by recognizing and detecting human emotions from text and images, or by trying to synthesize feelings and emotions in artificial creatures, like synthetic characters in computer games and robots.

Bibliography

Dodd, W. & Gutierrez R. (2005). The role of episodic memory and emotion in a cognitive robot. In: *Proceedings of the 14th Annual IEEE International Workshop on Robot and Human Interactive Communication.* Nashville: IEEE Press, pp. 692-697.

Dorneich, M.C.; Ververs, P.M.; Mathan, S. & Whitlow, S.D. (2005). A joint human-automation cognitive system to support rapid decision-making in hostile environments. In: *Proceedings of the 2005 IEEE International Conference on Systems, Man & Cybernetics.* Waikoloa (HI): IEEE Press, vol. 3, pp. 2390-2395.

Foteinos, V.; Kelaidonis, D.; Poulios, G.; Vlacheas, P.; Stavroulaki, V. & Demestichas, P. (2013). Cognitive management for the Internet of things: a framework for enabling autonomous applications. *IEEE Vehicular Technology Magazine,* 8(4), 90-99.

Gaussier P. (2001). Toward a cognitive system algebra: a perception/action perspective. In: *Proceedings of the European Workshop on Learning Robots (EWRL),* Prague, pp. 88-100.

Hauser, M.D.; Chomsky, N. & Fitch, W.T. (2002). The faculty of language: what is it, who has it, and how did it evolve? *Science,* 298 (5598), 1569-79.

Hershberg U. & Efroni, S. (2001). The immune system and other cognitive systems. *Complexity,* 6(5), 14-21.

Pattee, H. (1989). Simulations, realizations, and theories of life. In: Langton, C.G. (ed.). *Artificial life: proceedings of an interdisciplinary workshop on the synthesis and simulation of living systems.* Santa Fe Institute Studies in the Sciences of Complexity, 6. Redwood City (CA): Addison-Wesley, pp. 63-77.

Riegler, A. (2002). When is a cognitive system embodied? *Cognitive Systems Research,* 3(3), 339-348.

Varela, F.J.; Rosch, E. & Thompson, E. (1991). *The embodied mind: cognitive science and human experience.* Cambridge (MA): MIT Press.

Vlacheas P.; Giaffreda R.; Stavroulaki, V.; Kelaidonis, D.; Foteinos, V.; Poulios, G.; Demestichas, P.; Somov, A.; Biswas, A.R. & Moessner, K. (2013). Enabling smart cities through a cognitive management framework for the internet of things. *IEEE Communications Magazine,* 51(6), 102-111.

Xu, W.; Pang, J. & Luo, S. (2014). A novel cognitive system model and approach to transformation of information granules. *International Journal of Approximate Reasoning,* 55(3), 853-866.

Chapter 19

What is the relation between spike-timing-dependent plasticity and memory?

Fabio M. Simões-de-Souza

Center of Mathematics, Computation and Cognition (CMCC)

Federal University of ABC, São Bernardo do Campo, SP, Brazil

Fabio.souza@ufabc.edu.br

Gabriela Antunes

Laboratory of Neural Systems (SisNe), Department of Physics, FFCLRP

University de São Paulo, Ribeirão Preto, SP, Brazil

Abstract

This note summarizes the state of art of research on computer models of spike-timing-dependent plasticity (STDP), an important mechanism of long-term synaptic plasticity, essential for learning and memory.

Long-term synaptic plasticity is widespread in the brain and is considered the molecular and cellular substrate of learning and memory (Koekkoek et al., 2003; Schonewille et al., 2011; Whitlock et al., 2006). Long-term synaptic plasticity is expressed by a persistent change in the efficacy of synaptic transmission between neurons (Lüscher & Malenka, 2012; Lüscher et al., 1999). Respectively, long-term potentiation (LTP) and long-term depression (LTD) are defined by a persistent increase and decrease in the efficacy of synaptic transmission through changes in the number of AMPA receptors (AMPARs) in the postsynaptic membrane (Lüscher et al., 1999).

A special type of synaptic plasticity, called *spike-timing-dependent plasticity* (STDP), regulates the induction of LTP or LTD through the instant of occurrence of spikes (Bi & Poo, 2001). LTP occurs after the presynaptic neuron fires often, before the postsynaptic neuron fires. LTD happens after the postsynaptic neuron repeatedly fires earlier than the presynaptic neuron.

Many phenomenological models have been proposed to describe STDP. Abbott and colleagues (Song & Abbott, 2001; Abbott & Nelson, 2000) have proposed a simple model with only two differential equations that tracks the occurrence of pre and post synaptic spikes and explains satisfactorily the basic characteristics of STDP. However, more complex situations with multiple spikes require more complex models. For instance, by using triplets of spikes, little or no change occurs in the synaptic efficacy when a presynaptic spike is followed by a postsynaptic spike that is followed by another presynaptic spike (pre-post-pre). On the other hand, LTP happens after post-pre-post firing sequences. The triplet phenomenon was modeled with two differential equations with a suppression mechanism between spikes (Froemke & Dan, 2002). Later, the triplet phenomenon was accurately modeled with four differential equations (Pfister & Gerstner, 2006). However, multi-spike scenarios are complex and STDP responses to sequences of several spikes with Poisson distribution might not be explained precisely by actual STDP models. Parameter tuning can be used to induce different outcomes in STDP responses in simulations of multi-spike scenarios (Babadi & Abbott, 2013). Also, there are efforts to relate STDP to metaplasticity rules, such as the Bienenstock-Cooper-Munro (BCM) rule (Pfister & Gerstner, 2006; Izhikevich & Desai, 2003). The BCM rule is a firing rate based synaptic plasticity rule that robustly stabilizes bidirectional long-term synaptic plasticity (Cooper & Bear, 2012).

Phenomenological models of STDP are useful tools to replicate STDP responses, but they are limited and do not explain the intrinsic mechanisms underlying the occurrence of LTP and LTD. Presumably, STDP is determined by transient increases in the intracellular calcium ion concentration [Ca^{2+}] (Nevian & Sakmann, 2006). In this way, other models have been incorporating mechanisms of Ca^{2+} regulation (Rubin, 2005; Bartol et al., 2015) and activation of Ca^{2+}-dependent signaling pathways (Graupner & Brunel, 2007; Antunes et al., 2016a, 2016b). Also, recent evidence points to the role of metabotropic glutamate receptors (mGluR1) in the induction of STDP (Tigaret et al., 2016).

In addition, multiple forms of synaptic plasticity may coexist in neurons (Abbott & Nelson, 2000), and phenomenological models of STDP alone have not been sufficient to stabilize neural network models of assemblies of recurrent firing neurons performing associative learning and memory functions (Zenke et al., 2015). Thus, diverse synaptic plasticity mechanisms, including plasticity of inhibitory synapses, are required to stabilize memory networks of firing neurons (Zenke et al., 2015). However, STDP is not yet fully understood, and these limitations may rely simply on assumptions made by the actual simplified STDP models that do not fully explain the intracellular signaling networks that regulate AMPAR trafficking in the postsynaptic membrane. Certainly, more computational modeling efforts are required to create a solid ground to understand the relationship between in vivo multi-spike STDP and memory formation in large scale neural networks.

Acknowledgement

GA's work is funded by the DFG/FAPESP grant (IRTG 1740/TRP 2015/50122-0) and by a FAPESP postdoctoral fellowship (2014/08481-0). FMSS's work is funded by the IBM/FAPESP grant (2016/18825-4).

Bibliography

Abbott, L.F. & Nelson, S.B. (2000). Synaptic plasticity: taming the beast. *Nature Neuroscience*, 3, 1178-1183.

Antunes, G.; Roque, A.C. & Simoes-de-Souza, F.M. (2016a). Modelling intracellular competition for calcium: kinetic and thermodynamic control of different molecular modes of signal decoding. *Scientific Reports*, 6, article 23730.

Antunes, G.; Roque, A.C. & Simoes-de-Souza, F.M. (2016b). Stochastic induction of long-term potentiation and long-term depression. *Scientific Reports*, 6, article 30899.

Babadi, B. & Abbott, L.F. (2013). Stability and competition in multi-spike models of spike-timing dependent plasticity. *PLOS Computational Biology*, 12(3), e1004750.

Bartol, T.M.; Keller, D.X.; Kinney, J.P.; Bajaj, C.L.; Harris, K.M.; Sejnowski, T.J. & Kennedy, M.B. (2015). Computational reconstitution of spine calcium transients from individual proteins. *Frontiers in Synaptic Neuroscience*, 7, article 17.

Bi, G. & Poo, M. (2001). Synaptic modification by correlated activity: Hebb's postulate revisited. *Annual Review of Neuroscience*, 24, 139-166.

Cooper, L.N. & Bear, M.F. (2012). The BCM theory of synapse modification at 30: interaction of theory with experiment. *Nature Reviews Neuroscience*, 13(11), 798-810.

Froemke, R.C. & Dan, Y. (2002). Spike-timing-dependent synaptic modification induced by natural spike trains. Nature, 416 (6879), 433-438.

Graupner, M. & Brunel, N. (2007). STDP in a bistable synapse model based on CaMKII and associated signaling pathways", *PLOS Computational Biology*, 3(11), e221.

Izhikevich, E.M. & Desai, N.S. (2003). Relating STDP to BCM. *Neural Computation*, 15(7), 1511-1523.

Koekkoek, S.K.; Hulscher, H.C.; Dortland, B.R.; Hensbroek, R.A.; Elgersma, Y.; Ruigrok, T.J. & De Zeeuw, C.I. (2003). Cerebellar LTD and learning-dependent timing of conditioned eyelid responses. *Science*, 301 (5640), 1736-1739.

Lüscher, C. & Malenka, R. C. (2012). NMDA receptor-dependent long-term potentiation and long-term depression (LTP/LTD). *Cold Spring Harbor Perspectives in Biology*, 4(6), a005710, 15 pp.

Lüscher, C.; Xia, H.; Beattie, E.C.; Carroll, R.C.; von Zastrow, M.; Malenka, R.C. & Nicoll, R. A. (1999). Role of AMPA receptor cycling in synaptic transmission and plasticity. *Neuron*, 24(3), 649-658.

Nevian, T. & Sakmann, B. (2006). Spine Ca^{2+} signaling in spike-timing-dependent plasticity. *Journal of Neuroscience*, 26(43), 11001-11013.

Pfister, J.-P. & Gerstner, W. (2006). Triplets of spikes in a model of spike timing-dependent plasticity, *Journal of Neuroscience*, 26(38), 9673-9682.

Rubin, J.E. (2005). Calcium time course as a signal for spike-timing-dependent plasticity. *Journal of Neurophysiology*, 93(5), 2600-2613.

Schonewille, M.; Gao, Z.; Boele, H.J.; Veloz, M.F.; Amerika, W.E.; Simek, A.A.; De Jeu, M.T.; Steinberg, J.P.; Takamiya, K.; Hoebeek, F.E.; Linden, D.J.; Huganir, R.L. & De Zeeuw, C.I. (2011). Reevaluating the role of LTD in cerebellar motor learning. *Neuron*, 70(1), 43-50.

Song, S. & Abbott, L.F. (2001). Cortical development and remapping through spike timing-dependent plasticity. *Neuron*, 32(2), 339-350.

Tigaret, C.M.; Olivo, V.; Sadowski, J.H.L.P.; Ashby, M.C. & Mellor, J.R. (2016). Coordinated activation of distinct Ca^{2+} sources and metabotropic glutamate receptors encodes Hebbian synaptic plasticity. *Nature Communications*, 7, article 10289.

Whitlock, J.R.; Heynen, A.J.; Shuler, M.G. & Bear, M.F. (2006). Learning induces long-term potentiation in the hippocampus. *Science*, 313 (5790), 1093-1097.

Zenke, F.; Agnes, E.J. & Gerstner, W. (2015). Diverse synaptic plasticity mechanisms orchestrated to form and retrieve memories in spiking neural networks. *Nature Communications*, 6, article 6922.

Chapter 20

History of neuroscience and current philosophical problems

Francisco Rômulo Monte Ferreira
Graduate Program in Neuroscience and Behavior, Institute of Psychology
University of São Paulo (USP), São Paulo, SP, Brazil
fromulomonte@gmail.com

Maria Inês Nogueira
Laboratory of Neuroscience, Department of Anatomy
Inst. of Biomedical Sciences, University of São Paulo (USP), São Paulo, SP, Brazil
minog@usp.br

Abstract

Studies in the history of neuroscience give rise to interesting philosophical and methodological issues. In the first part of the paper, problems related to the formation and development of scientific concepts are examined, emphasizing how concepts such as "nervous cell" change with time. A distinction is drawn between concepts that have a clear referent from those that don't. In the second part, some methodological and philosophical issues of current research in neuroscience are examined.

1. The formation of concepts in neuroscience

In this discussion, one examines how the historical development of concepts may directly interfere in the research in Neuroscience. Reference will be made to the historical case studies upon which the text is inspired, but there is no space for a detailed description of them. We will start with a brief digression on the nature of scientific concepts and their connection to historical processes in the development of science.

1.1. Concept formation in science

Concepts in science are framed by the rationale of historical development, which cannot be understood solely from the textbooks which are used in the scientist's education or from the dynamics of experimental research.

Early in their education, scientists learn that concepts may be clarified by definitions. For example, the definition of "homeostasis" taught to undergraduate students in biological science courses is only somewhat related to the variations that such a concept has undergone before and after its formulation by Claude Bernard ([1878] 1974). The definition of "homeostasis" (in dictionaries or in textbooks of physiology) imposes a certain fixity which does not reflect the variation in time of the meaning of the term; this temporary fixity is what assures that the notion of homeostasis will be useful in scientific research, so that scientists don't have to continuously reformulate it. The *concept of homeostasis*, in opposition to its present definition, should be analyzed historically in terms of how its referent (in the case of an anatomic concept, for example) and its relations with the world (within the scopes of physiology, ontogenetic development, phylogenetic development, etc.) were formulated and modified. In these terms, the present definition of homeostasis is a subset of the concept of homeostasis, and such a concept can only be understood by following its historical formation and modification in the history of physiology.

A concept is a linguistic entity linked to a word without, however, being restricted to it. It is not an abstract entity, without any connection with experimentation (Koselleck, 2004). A scientific concept goes beyond its definition, to the extent that, in order to understand it, one must analyze historically its formulation (this does not mean that it is enough to register the first use of a word). Some features may be singled out in the definition of a concept, as in the following list of some properties that a concept possesses.

(1) *Invariance.* A concept represents a universal class which, in turn, may be instantiated by specific things or may be divided into smaller classes. Invariance basically occurs in three dimensions. (1.1) The psychological dimension, in which invariance comprises a mental state. For example, the invariance of the concept of "house" is the mental representation we have of a house. (1.2) The metaphysical dimension, where invariance is taken to be a substance or the fact that a concept represents a referent in the world. The ontology of this substance is formed in the physical or the transcendental world, depending on the metaphysical position adopted. (1.3) The linguistic dimension, where invariance is observed in language, as a derivation of social norms implied in language rules.

(2) *Criteria or rules* which enable us to identify and subsequently classify members of a certain category.

(3) The *manner* in which the formulation of a concept occurs. For example, certain concepts are initially thought as having a referent, such as the concept of a

"nerve cell". In its first formulation, there was the idea of "cell" and, consequently, "nervous cell" was taken to refer to the units that form the nervous tissue. But there are also concepts that don't have an anatomic referent, for example, psychological concepts, such as the concepts of emotion, pain, passion, among others, which don't have an anatomic referent in their original formulation. Instead, they have a behavioral referent. This is an important distinction, because in this case a concept is formulated taking the organism as a whole, and not considering one of its parts, as in our prior example, the cell. Cognitive neuroscience operates between these two types of concepts.

(4) *Organization.* The relations between concepts are important for understanding such concepts.

(5) The functions and relations of concepts *with the world.* A purely linguistic approach might assume only relationships between concepts, but this is not what is usually done in science.

1.2 Brief history of the concept of nervous cell

The formation of the concept of neuron in the 19th century marks the theoretical unification of the nervous system studies. From a theoretical perspective, we are able to assert that Neuroscience originated, in contemporary times, from the commonly called neuronal theory (Ferreira, 2013). The issue regarding the constitution of the nervous system was discussed by histologists during the major part of the 19th century. Based on the work of Luigi Galvani (1737-1798) and on an entire generation of researchers who examined the phenomena of bioelectricity at the end of the 18th century, the models for the nervous system from then on were thought of using as analogy the electric models, and no longer the mechanical models.

There was knowledge on the associations between fibers which grew from cellular bodies, and there were assumptions about their respective functions. Another characteristic of the nerve cells resulting from this discussion is the fact that the nerve conduction (the term "nerve force" was commonly used not only by histologists, but also by Charles Darwin and Richard Owen, who used this term in several moments) goes only in one direction from a cellular unit to another. The direction of the conduction was a problem both in the reticulated model of the nervous system (we will discuss this later on) and in the contiguous model, since even in the formation of a network we are able to identify cellular bodies.

"Dendrites" and "axons" are terms designating the extensions of the cellular bodies, and were introduced by Wilhelm His in 1890 and Albrecht Von Kölliker in 1896, respectively. The term "neuron" was first used in 1891 by Wilhelm Waldeyer and the term "synapse" was coined by Charles Sherrington in 1897, describing a hypothetical junction (Sherrington, 1948).

According to Finger (1994), three elements were needed to investigate the nervous system in the 19th century. (1) Improved techniques in microscopes which enabled visualizations with wide enlargements and no distortions. (2) Histological techniques that mark the cellular bodies and make it easier to visualize cellular processes. (3) The willingness of scientists to observe the slides with 'an open mind' to possible alternatives to the model in force at the time (the reticularist model). These three conditions were formed across the 19th century, and only in the second half we were able to identify their converging points.

A highly propagated idea was that the nerve cells are connected through anastomosis (merging), similar to parts of a tubulation. Deiters considered that the endings of the principal axis (axon) of a cell and the dendrites of the subsequent cell merged. These considerations were nothing but speculations, since Deiters's works do not have any observations that confirm such explanation.

Finger's hypothesis for the diffusion of continuous explanations (as opposed to contiguous models) for the nervous system have more to do with a research program which questions the localization theories developed in the first half of the 19th century rather than with actual empirical evidence. The reticulated model, according to its supporters, could better explain the quick transmission in the nervous system.

Albrecht von Kölliker (1817-1905), the supreme authority of histology at the time, postulated that only the dendrites of neighboring cells merged with other cellular endings (Kölliker, 1868). Subsequently, his postulation was expanded, assuming that there were other variations, axons merging with dendrites and with other axons.

Gerlach was an enthusiast of the ideas on merging in the nervous system, defended by Kölliker, and such idea led to thinking that the nerve impulse happened from cell to cell in fiber networks or as lattices. Accordingly, the nerve tissue would constitute itself as a network; a reticulum composed of a great number of physically interconnected parts.

In the mid-1860's, a period which strongly supported reticular theory, it was easy to accept Kölliker's and Gerlach's theory: they were authorities on the subject, and there was no empirical evidence which denied such model. Things would begin to change with a new method of tissue stain, the black reaction (*la reazione nera*), a method developed by Camillo Golgi (1843-1926). This method developed by Golgi would leave the cells with a black stain against a yellow background.

The proposition that the nerve cells would not form a network was made by renowned researchers. During the 1880's most experimental results corroborated the theory, but it was a professor from Madrid, who contributed the most to support the designated neuronal theory. Santiago Ramón y Cajal (1852-1934), a Spanish doctor and histologist, won the 1906 Nobel Prize for his work related to the basic unit of the nerve system, jointly with Golgi. His main works were about the thin structure of the central nervous system. Ramón y Cajal used the stain technique developed by Golgi with some changes, using silver chromate to stain some brain cells, in particular the dendritic trees and axons (extensions of the nerve cell).

Ramón y Cajal reached a very different conclusion from that of the reticularists; according to him, the nervous system is composed of billions of nerve cells (neurons) which are distinct and polarized. Ramón y Cajal suggested that instead of forming a network, neurons communicate with one another through a specialized mechanism (the synapse, although he did not coin the term). The postulation of the neurons as a basic unit of the nervous system is the basis of the neuronal theory, advocated by Ramón y Cajal as of 1888 (Ramón y Cajal, 1952).

In addition to using the method developed by Golgi, Ramón y Cajal studied the nerve tissue of newborn animals, which would enable observing the formation of neural circuits. We may simplify the proposed neuron doctrine – especially by Ramón y Cajal – to four general principles, i.e., there are four aspects that better define the concept of a neuron, currently: (1) The commonly called law of pericellular contact: the existence of individualized cells, the neurons. (2) The physical spacing between two cells, where there is communication between them, the synapse. (3) The neurotrophic law of neurogenesis, which enunciates that neuroblasts are guided towards chemical currents in the direction of secreting cells. (4) The law of dynamic polarization, which determines that information only goes one way.

The fourth principle represented, according to neuroscientist Eric Kandel, an extremely important aspect, because it enabled relating all the components of a nerve cell to only one function, namely, signaling.

> *It was not until 1955 that Cajal's intuitions were borne out conclusively. Sanford Palay and George Palade at the Rockefeller Institute used the electron microscope to demonstrate that in the vast majority of cases, a slight space – the synaptic cleft – separates the presynaptic terminal of one cell from the dendrite of another cell. Those new images also revealed that the synapse is asymmetrical, and that the machinery for releasing chemical transmitters, discovered much later, is located only in the presynaptic cell. This explains why information in a neural circuit flows in just one direction.* (Kandel, 2006, p. 69)

One of the greatest issues derived from the studies on neurons in the first half of the 20th century was how nerve cells communicated. The development of techniques to measure small currents was essential for these studies.

The form of the signal and its role in the codification of information would be addressed by Edgar Douglas Adrian (1889-1977) in the 1920's. After Adrian's work, new questions arose. How would a sensory neuron inform the intensity of a stimulus in the case of a light or deep touch, if a certain light radiates or blurs? How does it signal the duration of stimulus? In general terms, how do neurons differentiate a type of sensory information from another? (Kandel, 2006, pp. 77-78).

Adrian discovered that intensity results from the frequency in which action potentials are emitted (Adrian, 1931). Under a similar line of thinking, we may infer that the duration of a sensation is determined by the extension of time in which action potentials are generated. Another important scientist in this period was Charles Scott Sherrington (1857-1952), who discovered that not all nerve action is excitatory. He is commonly recognized for having explained what a synapse is.

Adrian and Sherrington were unarguably fundamental to unpuzzle nerve conduction. However, there were still many questions to be solved. What underlies the nervous system's notable capacity to conduct electricity in such a radical manner? How do electric signals connect and disconnect, and through which mechanism does propagation occur across the axon? A currently important work which was not mentioned, before due to exposition purposes, was the membrane hypothesis proposed by Julius Bernstein (1839-1917) around 1902. Bernstein, who was Helmholtz's student, realized that even when there was no neural activity, there was a difference in potential between the membrane and the external medium, and such difference in potential is what we call nowadays the resting potential of the membrane.

Kandel believes that Bernstein's membrane theory had an important role not only in the studies about signaling, but also in serving as a benchmark in the explanations of physicalists about brain functioning and in the then agonizing explanations of vitalists. Regardless of the definition, given in the 20th century, of the resting potential of the membrane by Bernstein, measuring small currents and the difference in potential in sections of animal bodies was a major problem for the studies about nerve conduction. We will discuss this further, to analyze some advances in what we consider the great experimental framework for the emergence of Hodgkin & Huxley in the 1940's and 1950's, with their ionic hypothesis.

1.3 Reification of concepts in cognitive neuroscience

In the formation of a specific research agenda for the Neurosciences in the 20th century, from the theoretical unity gained from the development of neuronal theory (the neuron concept) and from the further development of the disciplinary program of Neuroscience, the various research lines of this new scientific discipline had to adjust to the fundamental concepts that gave it theoretical support. Further awareness of the problem of cell signaling in the first half of the 20th century is a direct consequence of how fruitful the neuronal research program became.

Research on the neuron concept based on anatomy, physiology, embryology, etc., contributed to the development of the studies on the nervous system by the biomedical sciences without major problems. The studies on the then rising Cognitive Neuroscience constituted a separate case in the first half of the 20th century. Many concepts explored by cognitive neuroscientists refer to psychological attributes (emotion, thought, pain perception, among others). This study will now distinguish two modalities of concepts in which Neuroscience acts.

(*i*) Concepts with a clear referent (such as concepts within biological and biomedical disciplines). The concepts of neuron and cellular communication are examples of concepts with a clear referent. The former has a structural primary referent, and the latter has a functional referent (a set of operations which define the exchange of information between the cells of the nerve tissue).

(*ii*) Concepts without a clear referent in their first historical formulations (such as concepts explored by the Cognitive Sciences). The concept of emotion is an example of a concept without a clear referent, at least in the first modern formulations (considering examples in the 19th century). A clear referent in this case would correspond to parts or relations of an organism, instead of properties of an organism as a whole.

What is the relevance to the history of Neuroscience, or even to experimental Neuroscience, of such distinction between concepts? The first category of concepts (with a clear referent) does not constitute major problems in its historical development, since it is formed within the respective science. The modern concept of nerve cell carries in its genesis reference to the smallest morphological unit of the organic tissue, despite the fact that in the last hundred years elements of the neuron concept have changed. There is no problem of mereological order (an organism's part to whole relation). The second category of concepts (without a clear referent) refers to psychological attributes that in their origin do not show characteristics or relationships of parts of an organism, but rather show the organism as a whole.

The study of this topic constitutes a preliminary methodology proposal for research in the History of Neuroscience, focused on the historical formation of concepts. Cognitive Neuroscience in the 20th century, due to the development of the neuronal research program, and especially in order to establish experimental research protocols, acted toward reifying or embodying such concepts. Embodying these concepts is not a problem per se, since experimental results corroborate them with greater frequency. What I claim as relevant to the historical research on the formation of concepts in Neuroscience is indicating how these two categories of concepts are formed and how they were both standardized around the concept of neuron (the origin of the neuronal program). One problem that research in Cognitive Neuroscience will face in the future concerns the use of these concepts without a clear referent of a reified form in the creation of experimental protocols. This may lead to mereological problems, where psychological characteristics are attributed to specific parts of an organism (of the nervous system or even the brain).

2. Methodological and philosophical problems of neuroscience

2.1 Methodological issues

Imaging technology, besides X-rays, allowed for non-invasive approaches to the living and functioning brain in real time, and brought a revolution to medical diagnoses as well as to research. This technology enabled assessing the level of activity of different regions involved in the response to certain stimuli or in the condition of analysis. Among these methods, there is functional magnetic resonance imaging (fMRI) and positron emission tomography (PET). In such techniques, the convention that has been adopted is that the more intense the activity of a region in the brain of a living organism, the redder its color; as the activity weakens, the color goes from red to orange, yellow, blue and black.

These studies showed that, in fact, in a specific action, several cortical areas are involved, which favors the idea of non-localization of the function, involving the integrated action of neurons and glial cells of different regions, operating as a network or as neural circuits, in the execution of a given task, no matter how simple it may seem.

It is fascinating to observe that the brain continuously adapts to the circumstances of internal or external mediums, also altering its response potential. Its functioning is also influenced by modifications in the structure of a neuron or glia, by the chemical substances which surround them or which they produce, and by changes in the flow of substances which irrigate them or which are drained from them. Sensory stimuli may also modulate its action, whether through the birth of new cells (neurogenesis or gliogenesis), the programed death of cells, the formation of new synapses or their degeneration (Lent, 2010; Motta-Teixeira et al., 2016).

Recently, a breakthrough brain map was originated from advanced scanners and computers running artificial intelligence programs which were prepared to identify the brain's hidden regions, from vast amounts of data collected from hundreds of test subjects who participated in the Human Connectome Project (Zimmer, 2016). This new atlas is great achievement, since 97 new brain regions were added to the 83 already known, which were also mapped. This collected data revealed not only the arrangement of the neural network, but activity and myelin insulation in different areas, among 112 registered characteristics still in analysis.

These results emphasize that the brain is composed of various types of cells, forming stations and networks of information processing, to the extent that localizationism and holism coexist, but in a manner that is different from the traditional view. In spite of this evolution of morphofunctional understanding, fundamental questions about the mind and consciousness still remain.

Silva & Bickle (2009, pp. 91-93) propose a new discipline to deal with the understanding of science nowadays, the "science of research", especially applied to (1) neurobiology and its molecular approach to cognition and (2) the inefficiencies in institutionalized science. About the first point, they write:

> *How do recent proponents of mind-to-molecular pathways linkages justify their claims? What kinds of experimental results do they take as evidence for such claims, and when do they claim to have enough (sufficient) evidence? Lurking just below the surface of this puzzle is the broader methodological question of how we discover and test answers to such questions.*

Comparing the brain to a machine makes reference to cognitive science, computers and artificial intelligence, which seek to reproduce the functioning of the brain, correlating it with the mind. Research in these fields raises the question of what is human nature, but they also lead to robotics and the possible extinction of human nature as we know it.

In relation to the brain-machine interaction, cybernetics or bionics, Miguel Nicolelis, a Brazilian physician and researcher at Duke University, believes in the potential of technology to transform society and shape a promising and innovative "industry of the brain". Nicolelis, in addition to exploring paths to cure or minimize neurologic disorders such as Parkinson's and Alzheimer's diseases, has also been working on an exoskeleton linked to a computational system for helping paraplegic people. He believes neuroscience will be important for a better and fairer society, and contends that a more holistic approach to the understanding of the brain will be necessary (Nicolelis, 2011).

2.2 Philosophical debates in cognitive neuroscience

Reason and emotion do not constitute a dichotomy, but rather act in synergy with decision-making. The nature of emotions is explained as physiological and psychological reactions that influence perception, learning and performance, and they may be innate or learned. Accordingly, Damasio defines emotion as certain organic alterations, which leads to a subjective "feeling". This idea, that a bodily reaction causes a feeling, is based on the theory of emotions of William James (1890) and C.G. Lange (1922). Damasio developed the hypothesis of the "somatic marker", which are the bodily responses to images (or thoughts), associated to feelings that arise from emotions, and may be involved in the process of decision making among a certain array of options. The influence of the marker might be positive or negative. "When a negative somatic marker is juxtaposed to a particular future outcome the combination functions as an alarm bell. When a positive somatic marker is juxtaposed instead, it becomes a beacon of incentive." (Damasio, 1994, p. 174).

Max Bennett, a neuroscientist, and Peter Hacker, a philosopher, criticize the mainstream of neuroscience, arguing that language is being misused, based on the principles of ordinary Language Philosophy, inspired by the philosopher Ludwig Wittgenstein and Gilbert Ryle (Bennett & Hacker, 2003). They thus point out what they consider confused and incoherent results and concepts of leading researchers in the field of cognitive neuroscience. For example, they advocate that only a complete human being can have a "mind", when referring to the Gazzaniga's and LeDoux's experiments on commissurotomy (brains split by severing the corpus callosum) and their conclusion is that each of the hemispheres of the split brain have a mind of itself. In spite of the strong reactions that the book raised from several neuro- and cognitive scientists, the book's major contribution is to open a debate between scientists and philosophers.

Their second book focused on the history of cognitive neuroscience, and although they maintain their point of view, they establish a more comprehensive dialogue with their opponents. They hope not only to answer the criticisms made to them, but also continue their attacks on the conceptual presuppositions of cognitive neuroscience, fostering a critical analysis of research experiments and interpretations in favor of a true contribution to knowledge. Concerning the views of Damasio, Bennett & Hacker (2013, p. 196) claim that:

> *Damasio's theory inherits the defects of its Jamesian ancestor. There are extensive conceptual confusions in his somatic marker hypothesis [...] to have or feel an emotion typically involves a variety of cognitive components – but having mental images is not a necessary accompaniment of having an emotion; nor is it uniformly a cause of any*

> *somatic perturbation that may be involved in the emotional episode.*

This is Bennett & Hacker's first objection to Damasio's theory, among nine of them. The second is that these authors understand that, contrary to Damasio, emotions and feelings are two sides of the same coin, and this error they attribute to the third objection, i.e., Damasio's failure to distinguish between the causes of an emotion and its object.

Another general criticism to mainstream cognitive neuroscience is made by Maturana & Varela (1992, p. 169):

> *[The nervous system] is the result of a phylogenetic drift of unities centered on their own dynamics of states. [...] the nervous system does not "pick up information" from the environment, as we often hear. On the contrary, it brings forth a world by specifying what patterns of the environment are perturbations and what changes trigger them in the organism.*

According to these authors, the nervous system, the rest of the body, and the environment interact, exert reciprocal influence; a particular nervous system results from its phylo-ontogenetic evolutionary process, from experiences with its surroundings, which influence the construction of knowledge. The nervous system participates in cognitive phenomena in two complementary ways: the first is through the expansion of his sensorial capacities, while the second is an expansion of its internal states (Maturana & Varela, 1992, p. 175).

The philosophical debate on cognition and the mind-body problem continues with many different opinions. In a recent conference, where most participants could be considered materialists, in the sense that there is no evidential support for a nonphysical egos or soul, and emergentists, there still were many diverse theories (Johnson, 2016). Neuroscientist Michael Graziano claims that the brain is like an evolved computer that simulates the outside world, and within this simulation it represents itself, thus creating an illusion of the self, so the machine mistakenly thinks it has "magic" inside itself. The old doctrine of panpsychism has been receiving much attention, in its view that all of nature has mentality. Physicist Max Tegmark, has proposed a new state of matter, "perceptronium", which can process information and give rise to subjectivity. On this he was inspired by neuroscientist Giulio Tononi and his theory of integrated information, which claims that simple devices that process information also have a low level of consciousness. The conclusion on what is the nature of cognition or consciousness is far from being attained.

From the neuroscientific side, it is worth mentioning the research which has counted in a precise way the number of brain cells in the human beings and in other animals (Fonseca-Azevedo & Herculano-Houzel, 2012), and the association of increase in brain size in evolution with the appearance of the increased energetic source provided by the cooked food diet (Carmody & Wrangham, 2009), something that we did not suspect some years ago.

To conclude, recall the movie *2001: A space odyssey*, directed by Stanley Kubrick, where human evolution would symbolically jump whenever a huge monolith appeared. Maybe we could consider the development of cooked food and the ensuing increase in human brain size as a driving force analogous to the monolith, besides the development of tools (shown in the film), which helped shape the nature of the Homo sapiens. What will be the next achievement in brain research that might invoke Kubrick's monolith, and crack open the scientific understanding of subjectivity?[1]

[1] Section 1 was written by Francisco Rômulo Monte Ferreira, and section 2 by Maria Inês Nogueira.

Bibliography

Adrian, E.D. (1931). *The mechanism of nervous action: electrical studies of the neurone.* Philadelphia: University of Pennsylvania Press.

Carmody, R.N. & Wrangham, R.W. (2009). The energetic significance of cooking. *Journal of Human Evolution*, 57(4), 379-391.

Bennett, M.R. & Hacker, P.M.S. (2003). *Philosophical foundations of neuroscience.* Oxford: Blackwell.

——— (2013). *History of cognitive neuroscience.* Oxford: Wiley-Blackwell.

Bernard, C. ([1878] 1974). *Lectures on the phenomena common to animals and plants.* Transl. H.E. Hoff, R. Guillemin & L. Guillemin. Springfield (IL): Charles C. Thomas.

Damasio, A.R. (1994). *Descartes' error: emotion, reason, and the human brain.* New York: Avon.

Ferreira, F.R.M. (2013). *A teoria neuronal de Santiago Ramón y Cajal.* Ph.D. thesis. São Paulo: University of São Paulo.

Finger, S. (1994). *Origins of neuroscience: a history of explorations into brain function.* New York: Oxford University Press.

Fonseca-Azevedo K. & Herculano-Houzel S. (2012). Metabolic constraint imposes tradeoff between body size and number of brain neurons in human evolution. *Proceedings of the National Academy of Science USA*, 109(45), 18571-18576.

Johnson, G. (2016). Consciousness: the mind messing with the mind. *New York Times*, July 4. Available at: https://www.nytimes.com/2016/07/05/science/what-is-consciousness.html

Kandel, E.R. (2006). *In search of memory: the emergence of a new science of mind.* New York: W.W. Norton.

Kölliker, A.V. (1868). *Elements d'histologie humaine.* Paris: Masson.

Koselleck, R. (2004). *Futures past: on the semantics of historical time.* Transl. K. Tribe. New York: Columbia University Press. German original: 1979.

Lent, R. (2010). *Cem bilhões de neurônios? Conceitos fundamentais de neurociência.* Rio de Janeiro: Atheneu.

Maturana, H.R. & Varela, F.J. (1992). *The tree of knowledge: the biological roots of human understanding.* Transl. R. Paolucci. Boston: Shambhala.

Motta-Teixeira, L.C.; Takada, S.H.; Machado-Nils, A.V.; Nogueira, M.I.; Xavier, G.F. (2016). Spatial learning and neurogenesis: effects of cessation of wheel running and survival of novel neurons by engagement in cognitive tasks. *Hippocampus*, 26(6), 794-803.

Nicolelis, M. (2011). *Beyond boundaries: the new neuroscience of connecting brains with machines – and how it will change our lives.* New York: Macmillan.

Ramón y Cajal, S. (1952) *¿Neuronismo o reticularismo? Las pruebas objetivas de la unidad anatomica de las celulas nerviosas.* Madrid: Instituto Cajal.

Sherrington, C. (1948). *The integrative action of the nervous system.* Cambridge (GB): Cambridge University Press.

Silva, A.J. & Bickle, J. (2009). The Science of Research and the search for molecular mechanisms of cognitive functions. In: Bickle, J. (ed.). *The Oxford handbook of philosophy and neuroscience.* New York: Oxford University Press, pp. 91-115.

Zimmer, C. (2016). Updated brain map identifies nearly 100 new regions. *New York Times*, July 20. Available at: http://nyti.ms/2a9aU2a.

Index

D

E

F

G

H

I

J

K

L

M

N

O

P

W

www.ingramcontent.com/pod-product-compliance
Lightning Source LLC
LaVergne TN
LVHW010722210926
848485LV00007B/1420/J